Dedicated to the late John Michell whose original CITY OF REVELATION, contained a huge amount of information that helped me understand the Earthstars Landscape Temple geometry through its connections to John's New Jerusalem diagram.

Without his work, knowledge, insight, vision and encouragement, the significance of many aspects of the discovery would have almost certainly remained totally unfathomable to my lesser intellect.

By the same author

EARTHSTARS. Hermitage Publishing (1990)

EARTHSTARS THE VISIONARY LANDSCAPE Hermitage Publishing (limited edition hardback 2000)

LONDON'S CAMELOT AND THE SECRETS OF THE GRAIL
Earthstars Publishing (2009)

LONDON'S LEY LINES – PATHWAYS OF ENLIGHTENMENT
Earthstars Publishing (2010)

THE BEER GURU'S GUIDE A path of enlightenment for those who thirst for knowledge   Souvenir Press   (2006)

See www.earthstars.co.uk for details.

The contents of this book were first published in 2000 by Hermitage Publishing, as **EARTHSTARS THE VISIONARY LANDSCAPE Part One, London City of Revelation.** This was a limited edition hardback print run of just 1,000 copies. Now out of print, rare and collectible.

This 2010 paperback by Earthstars Publishing is essentially a more affordable version of the same book.

Copyright; C.E.Street.
The moral right of the author has been asserted.
All rights reserved.

Ordnance Survey maps utilised in this book are based upon the O.S. 1:50,000 First Series and Landranger Series and are reproduced with permission of Her Majesty's Stationary Office. Crown Copyright. All rights reserved. License Number 10029558.

All maps are Crown copyright and may not be reproduced without prior permission of HMSO. Permission from Earthstars Publishing and the author would also be required.

No part of this publication may be reproduced, copied, stored in a retrieval system, or transmitted in any form, by any means, electronic, mechanical, photocopying or otherwise without prior consent of the copyright owner or publisher.

A CIP catalogue record for this book is available from The British Library.

ISBN: 978 0 9515967 5 3

Published by Earthstars Publishing, midsummer Solstice 2010

www.earthstars.co.uk
www.earthstars.co.uk

LONDON CITY OF REVELATION

Previously published as

Earthstars The Visionary Landscape

Christopher E. Street

Earthstars Publishing London N14 6LP.

LONDON CITY OF REVELATION

CONTENTS	Page
FOREWORD by John Michell	ix
THE VISIONARY LANDSCAPE	1
A WAKE UP CALL	9
LONDON'S LOST CAMELOT	15
The Mystic Isle	20
The Lady of the Well	23
Arthurian Mysteries	30
THE DREAM	33
St. Mary The Virgin, East Barnet	34
A Spiritual Power station?	39
The Barnet Triangle	42
Barnet Triangle Dimensions	43
HIDDEN CONNECTIONS	45
Green Park's Grove	52
Westminster Cathedral	54
London's Hilltop Henge	54
Long distance information	58
Lines of Spirit	60
The Croydon Triangles	64
Croydon Triangle Dimensions	67
A crop of circles	68
Circle dimensions	68
THE FIRST EARTHSTAR	70
Pentagonal mark points	74
The Horsenden Hill axis	74
The London Stone	77
St. Bride's, Fleet Street	78
Tower Hill	78
Horsenden Hill	79
One Tree Hill	81
The Burnt Oak to Bellingham axis	83
The Caesar's Camp axis	84
The Sion Park to Wanstead axis	86
Five-point star intersections	87
A CLUSTER OF EARTHSTARS	91
Sion Park	96
Watling Park	96
The eight-point star	96
The Camelot Pentagram	105
The eighteen-sided polyhedron	109
The Croydon Triangle star	112

CONTENTS continued   **Page**

THE DIVINE PLAN	114
Sacred Geometry	116
The Squared Circle	119
Harmonic numbers	126
THE FULL CIRCLES	130
East Ham Triangle	130
Hanwell Triangle	132
The 20-point Star	137
The 30-point Star	140
The patterns so far	149
PRACTICAL APPLICATIONS OF SACRED GEOMETRY	151
Building in the Spiritual Dimension	159
The Geometry of the Soul	169
THE WEB OF LIFE	170
As above, so below	178
The Horsenden Hill Midsummer Sunrise Line	179
The Coronation Line	182
The Stonehenge Line	186
Life's Rich Pattern	187
THE JOURNEY TO THE CENTRE	190
Microcosm and Macrocosm	196
OTHER STAR SYSTEMS	207
The Circle of Perpetual Choirs	207
David Wood's GENISIS	211
Henry Lincoln's THE HOLY PLACE	213
Pi in The Sky	215
On Earth as it is in Heaven	218
The Keys to the Temple	219
Other Earthstars Patterns	219
Sacred Geometry of Crop Circles	222
John Michell's Sacred Geometry of Old Jerusalem	223
There's more than one Pentagon in Washington	226
LONDON, CITY OF REVELATION	229
The Gates of the City	239
The Pearl of Heaven	240
THE BRIDE	246
The Lion and The Unicorn	253
THE END OF AN AGE ?	257
The Mayan Prophecies	259
The Rainbow Warriors	260
New age or Ancient wisdom ?	261
The Second Coming	263

CONTENTS continued   **Page**

THE WATERS OF LIFE   269
  The Fountain Effect   273

CAMELOT, THE ROUND TABLE AND THE HOLY GRAIL   277
Camelot and the New Jerusalem   283

THE EARTHSTARS VISION   285
Geosophy   287
Healing the Wasteland   289
The Earthstars Vision   293

Blake's Jerusalem   301

Bibliography   302

ILLUSTRATIONS

1; Stonehenge		10
2; 1658 Map showing Camlet Moat as Camelot		18
3; Plan of Camlet Moat		19
4; The NE corner of Camlet Moat		21
5; The remains of the well at Camlet Moat		24
6; A Virgin Mary illustration similar to the apparition of the White Lady		25
7; The Tarot High Priestess card		26
8; The Parish Church of St. Mary the Virgin, Monken Hadley		28
9; Charles Williams' map showing Camelot in North London		32
10; The Parish Church of St. Mary The Virgin, East Barnet		35
11; The Minchenden Oak alignment		37
12; The Primrose Hill /Parliament Hill/East Barnet alignment		38
13; The Empress of the Tarot		39
14; The Barnet Triangle		44
15; The Barnet Triangle extended lines		46
16; The Earthstars' main North-South alignment		50
17; St. Joseph's Church, Highgate		51
18; The Queen Victoria Memorial Fountain		51
19; Green Park's Grove		53
20; Westminster Cathedral		55
21; Pollard's Hill, Norbury		57
22; The North-South line well defined over a hundred miles		59
23; The well at Rottingdean		60
24; The Madonna and child statue at Westminster Cathedral		61
25; The Croydon Triangle		65
26; Croydon's isosceles triangle		66
27; The first Earthstar's circles		69
28; London's pentagonal Earthstar		71
29; London's main roads run parallel to the star's axes		73
30; Horsenden Hill axis diagram		74
31; Horsenden Hill axis map		76
32; Horsenden Hill, Greenford		79
33; The Major Arcana Tarot card of Strength		80
34; One tree Hill, Alperton		82
35 Diagram of the Burnt Oak to Bellingham axis		82
36; Diagram of the Caesar's Camp axis		84

37; St. Eloy's Well at Tottenham	85
38; Diagram of the Sion Park to Wanstead axis	86
39; The Inner pentagram	87
40; The pentagonal mark point on Tower Hill	88
41; Alexandra Palace	89
42; The huge Minchenden Oak near Christ Church, Southgate	90
43; Barnet Triangle extended line	92
44-45; Ten point star diagram	93
46; Ten point star map	95
47; Grass circle at Burnt Oak	97
48-49; 8 point star diagram	97
50; 8 point star map	97
51a and 51b; Finsbury Park stones	101
52; All Soul's Church, Langham Place	103
53; Silbury Hill	104
54; 8 point and 5 point star connection	104
55-56-57; Camelot pentagram diagrams	106
58; Camlet Moat /Bellingham church relationship	107
59; Camelot Pentagram map	108
60; The eighteen-sided polyhedron	111
61-62; Croydon triangle's 5 point star diagram	112
63; Croydon triangle's 5 point star map	113
64; Pentagonal symmetry common to many flowers	115
65; Starfish	115
66; Hexagonal structure of a snowflake	115
67; Hexagonal formation of Uranium atoms	115
68; Vesica Piscis	117
69; Vesica's hexagonal structure	118
70; Vesica's pentagonal structure	118
71; Squared circle diagram	120
72-73; Squared circle construction	121
74-76; Squared circle construction	122
17) Elemental points	124
78; Squared circle map	125
79; Planetary proportions in the squared circle diagram	126
80; The East Ham Triangle	131
81; The Hanwell Triangle Map	133
82-83; East and west 5point star diagrams	134

84; East Ham 5 point star map	135
85; Hanwell 3 point star map	136
86; 20 point star circle of the Earth Spirit	138
87; 20 point star diagram	139
88; The wheel of the year	139
89; The 30 point star map	141
90; 15 point star map	142
91; Shared axes of 5,10 and 30 point stars	144
92; The hexagram linking the 18 and 30 point stars	145
93 and 94; The 30 point star as 6 x 5 or 5 x 6	147
95; 30 point star's hexagonal axes	148
96; The six point star on the inner circle	149
97; The squared circle and The Great Pyramid	152
98; Stonehenge's Earthstar layout	153
99-100; Pentagonal and hexagonal patterns at Stonehenge	154
101; Stonehenge's 30 point star	155
102; Stonehenge's 20 point star	156
103-104; A 40 point star as 5 x 8 or 8 x 5	157
105; Egyptian rock tomb	159
106; King's College Chapel	160
107; Lady Chapel, Glastonbury	161
108; The common framework of Ad Triangulum and Ad Quadratum	162
109; Glastonbury Abbey ground plans	163
110; Gothic Standard Plan	164
111; Westminster Abbey	165
112-116; The earliest Temple Groundplan	167
117; Symbolic geometry of the seven chakras	172
118; The seven chakras	173
119; The Earthstars' pentagon's midsummer sunrise line	180
120; Midsummer sunrise at St. Mary's East Barnet	181
121; Map of the Westminster abbey/St. Paul's sunrise line	183
122; The ancient mound at Arnold Circus	184
123; The Kingstone at Kingston-on-Thames	184
124; The Barking Abbey/Westminster Abbey line	185

125; The Stonehenge Line	186
126; Full diagram of the 30-point star	187
127; Full diagram of the 20-point star	188
128; Map showing the St. Paul's Portland Chapel at the Earthstars' centre	192
129; Engraving of St. Paul's Chapel	193
130; Brock House	193
131-132; St Mary's Chapel, Glastonbury Abbey	195
133-134; Stars within stars	196
135; The inner 5 point star	198
136; Inner 5 point diagram and mark points	199
137; Map of the 5 point star within the inner 5 point	200
138; Westminster Abbey and St. Paul's circles map	202
139; Westminster Abbey and St. Paul's vesica piscis map	203
140; Westminster Abbey, St. Paul's and the Earthstars' centre	204
141; St. Martins/Westminster Abbey/Green park triangle	206
142; Arc of the circle of perpetual choirs	208
143; The circle of perpetual choirs	209
144; David Wood's Genisis landscape geometry	212
145 and 146; The Holy Place. Henry Lincoln's Rennes patterns	214
147; Dowsed Earth star patterns	216
148; Dowsed Earth star pattern	217
149; Greg Rigsby's pentagram of French Cathedrals	219
150; Pentagonal pattern of hilltops centred on Harrow on the Hill	220
151; Complex crop pattern near Stonehenge	221
152 and 153 ; Pentagonal crop circle patterns	221
154; Pentagonal geometry in Jerusalem	224
155; Pentagram in Washingon Street layout	227
156-162; 12 x 12 grid patterns	233-235
163-164; Squared circle constructions on a 12 x 12 grid	236
165; John Michell's New Jerusalem diagram	241
166; The Dove of Peace	244
167; The Lion	255
168; The Unicorn	256
169; The Tarot Star	270
170; The Rag Tree overhanging the well at Camlet Moat	282
171; The temple "folly" at Audley End House	282

FOREWORD
by
John Michell

Reproduced from the original 1990 Earthstars

The mystery of London goes back to the earliest times, for no-one knows when or how the city was founded. Its legendary history begins in about 1200B.C. when Brutus and an army of Trojans landed in Totnes and having defeated the native giants, occupied London and made it the capital of their realm. Their name for it, New Troy, was still current when Caesar invaded in 54 B.C. for the people he found in possession of London were called the Tri-Novantes or New Trojans.

The dynasty of Trojan rulers after Brutus is chronicled into the early Christian era. Every London child up to the time of Shakespeare was taught about those old kings and knew the places associated with them. On Ludgate Hill, named after King Lud who fortified London in about 70 B.C. stood the temple of Apollo on the site now occupied by St. Paul's Cathedral. All around it, in the traditional square-mile area of the city, were numerous shrines. Their sites were presumably located following the invariable practice of antiquity, by the state augurs or geomancers, who professed the art of creating a religious enchantment over cities and country-sides, and their original pattern was largely preserved by the succession of temples and churches built upon the same sites. It survived even the Great Fire of 1666 when Sir Christopher Wren rebuilt most of the old churches on the ruins of their predecessors. Before the fire there were more than a hundred churches, each with its tiny parish, within the city's square mile. In many cases, they have proved to have been located upon the prehistoric sanctuaries.

As London has developed from its ancient foundations, it has constantly received imprints from mystical architects and planners. The influence of masonic augury - a tradition closely guarded and passed down within the fraternity - is apparent in the alignments of medieval churches, such as The Strand Ley (St. Martin-in-the-Fields; St. Mary-le-Strand; St. Clement Danes; St Dunstan's Fleet Street to Arnold Circus) and the line of five churches, including St. Paul's between St. Clement Danes and St. Dunstan's Stepney. These, no doubt, were planned as part of a larger pattern, whose overall purpose was to attract divine influences and to procure good fortune and prosperity for the citizens of London.

Students of these matters dispute about the duration of the priestly tradition and the degree to which it has persisted in modern times. It is certainly apparent in the 17th and 18th centuries, when some of its principles and methods were applied by learned architects and landscape designers to the creation of miniature paradises in noblemen's estates. An esoteric influence has also been discerned in the works of certain 19th-century church architects. The influence persists to this day, not so much through the moribund arcana of masonry as through the autonomous nature of the tradition itself. Being rooted in human nature, through the constant laws of aesthetics which pertain also to the human soul, the tradition is self-renewing.

Every true artist can invoke it and bring its influence to bear on whatever they create. Many in all ages have done so. Chris Street's bold thesis of a sacred design, spanning modern London and extending beyond it, does not, therefore, imply a single human designer. London is the creation of thousands of different builders and architects, working over many centuries. Their common humanity has exposed them to a common influence, and thus together they have created a pattern of which not a single one of them may have been conscious. Chris has discovered that pattern. He has not just dreamt it up, for the fact of its existence is beautifully illustrated on his maps. Yet it is undoubtedly a dream pattern, detached from mundane reality and far transcending both the imagination and the capacity of any town-planning department.

In setting out his wondrous discoveries, Chris is exercising the truest and noblest function of a poet, translating images from the world of dreams and ideals into the reality of this present world.

It is impossible to read this honest and visionary book without being deeply affected. The process begins early on, as London's secret pattern is unfolded. The first reaction of the modern educated mind is to be sceptical, even affronted, at this apparent elaboration of nonsense. What are we being asked to believe? It is soon apparent that no particular beliefs are being asked for. Chris demonstrates his patterns. They are obviously attractive and as we look at them, we fall under a poetic spell. The critical, rationalizing aspect of our minds is not being challenged or tested. It is simply by-passed and we are immediately engaged on a higher level, through the imagination. On that level the mind is laid open to influence, and here the author takes on a serious responsibility. He could easily, at this stage, have deluded us with personal fantasies. yet in this book we are not trifled with, for the author is well equal to his responsibility, and under his guidance we follow the classical path through the studies which lead towards initiation.

This journey is naturally delightful, and its pleasures are much enhanced by the charm and intelligence of our guide. He has captured the essence of the ancient tradition. It lives within him and inspires his writing. This is poetry of the highest quality, serving the highest possible purpose. Here revealed is a world-view, quite different from the low-level, materialistic world-view which dominates modern institutions and evidently destined to replace it.

A transformation of perceptions, both personal and general, seems inevitable in these present times. Chris Street's book is a symptom of that transformation. It is not belittling him to say that it is one of those books that is greater than its human author. It has clearly been written under guidance. To the author's personal credit is that he had made himself worthy of transmitting the insights and revelations which he has invoked directly from the course of an ever-living tradition.

John Michell.

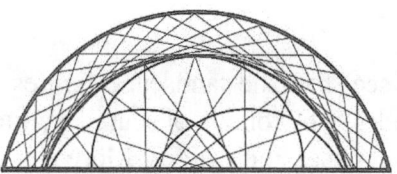

Introduction
The Visionary Landscape

All things are connected, or so every mystic tradition tells us.

Back in 1982, I began to get an inkling of what that meant, at least as far as some of London's ancient sacred sites were concerned. It was around that time that I first stumbled upon the discovery that they are not scattered about the capital at random. Their locations are far from arbitrary. Many are arranged in precise geometric patterns upon the landscape.

By no stretch of the imagination do these patterns seem to be a chance occurrence. Every single one of them connects to the rest to create a vast and beautiful design resembling an immense mandala covering the whole of Greater London, over 400 square miles. Nor does this seem to be an isolated configuration. The main alignments can be extended to link with ancient sites the length and breadth of the land.

Further evidence suggesting something other than a random formation is apparent in the geometric form of the design itself. It is a specific and recognisable construction of sacred geometry, a temple ground plan of immense significance. In fact, the design forms an integral part of the foundational geometry of several ancient sacred structures, including Stonehenge and the Pantheon in Rome.

The basis of the design, however, has far more mysterious origins and was drawn from something much older than even Stonehenge: the Earth itself. The proportional relationship shared by the radii of the two concentric Earthstars' circles corresponds directly to the proportional relationship shared by the radii of the Earth and Moon.

The Earthstars mandala contains many other significant planetary dimensions within its geometry. For instance, it is precisely 1080 times larger than Stonehenge and 1080 in miles is the mean radius of the Moon. Needless to say, all this adds up to an enigma of cosmic proportions and one which raises a great many questions about its origins, its nature and its purpose.

However, its high degree of intangibility makes it understandably difficult to evaluate. We are not dealing here with a pattern that is laid out clearly for all to see, like the landscape figures etched into the Nazca Plains high in the Andes. This one is obscured by miles of streets, housing estates, factories, warehouses and offices. Moreover, there are no real lines on the landscape. The patterns exist only by virtue of the sites that define them and the lines drawn on my maps. For some, that means they do not exist at all. End of story. Problem neatly swept under the carpet. It is not as simple as that.

This is a phenomenon that has its roots in our sacred sites: places that are supposed to have mysterious and intangible connections: places where we traditionally go to connect to our spiritual dimensions.

Very few people seem to take that notion literally these days, but there is considerable evidence to suggest that we should. It is widely accepted that our oldest sacred sites have a history of successive use that may well date back to humanity's earliest spiritual experiences. Long before a church or anything else was built upon these places, they were recognised as hallowed ground. Why?

What for instance, inspired our ancestors to regard Ludgate Hill as sacred when it was crowned only by grass and trees, rather than St Paul's Cathedral, or even the temple of Diana which reputedly preceded any church on the site ? A splendid view over miles of unspoilt countryside wouldn't have been a novelty in those days.

The answer lies not in the physical attributes of these locations, but in other, less obvious factors; the atmosphere; the sense of presence; the spirit of the place. These were places of power. Places of the gods. Places of vision. Places of healing.

The temples of ancient Greece are prime examples. They were not simply locations where people would congregate to worship their deities and perform religious ceremonies. By all accounts, many occupied a spot where the gods or goddesses of classical antiquity had actually been seen, in a visionary sense if not physically, or at the very least, had been contacted through an oracle. Any shrine or temple marking the spot was usually a later addition in honour of the appropriate deity or spirit of place.

The Temple of Athena at Lindos on Rhodes, for instance, stands directly above a cavern where a vision of the goddess is said to have actually appeared. The Temple of Apollo at Delphi on the slopes of Mount Parnassus arose from the site's use as the pre-Christian world's foremost oracular shrine. Other similar examples abound. Tales of mythic Gods, oracles, seers and visions are all too easily dismissed as fantasy by the sceptical, but similar events have continued to occur throughout the centuries, even to the present day. In a cavern at Lourdes in 1923, Bernadette Soubirou saw a vision of a "white lady" understood to be the Virgin Mary The grotto has been a healing shrine ever since, attracting thousands of pilgrims every year.

An appearance of the Virgin Mary at Walsingham in Norfolk, witnessed by Richeldis de Faverches, prompted her to commission the building of a shrine there in 1061.

In more recent times, at Fatima in Spain and Medjugorje in what was formerly known as Yugoslavia, visions of the Virgin appeared and even gave prophecies to some of the assembled witnesses. Sightings of the Virgin Mary were even reported at Willesden Green in London as recently as the 1970's. According to Michael S. Durham's book, MIRACLES OF MARY, there have been 21,000 recorded apparitions of the Virgin over the last ten centuries and many more unrecorded.

Obviously, an element of clairvoyance, involuntary or otherwise, may be involved, but generally, it seems to be particular places that trigger an experience, as well as people. In fact, the ability of certain locations to stimulate personal spiritual experiences, often of a visionary nature, is probably one of the reasons why our sacred sites came to be recognised as sacred in the first place. It is certainly a key factor in how the phenomenon of leys came about.

Alfred Watkins, the person generally held responsible for introducing the concept of leys to the public at large, maintained throughout his life that the entire discovery came to him in a single flash of inspiration, akin to a mystical vision. In the introduction to the Abacus reprint of Watkins' THE OLD STRAIGHT TRACK, John Michell describes it thus:

"The Revelation took place when Watkins was 65 years old. Riding across the hills near Bredwardine in his native county, he pulled up his horse to look out over the landscape below. At that moment, he became aware of a network of lines, standing out like glowing wires all over the surface of the country, intersecting the sites of churches, old stones and other spots of traditional sanctity. The vision is not recorded in The Old Straight Track, but throughout his life Watkins privately maintained that he had perceived the existence of the ley system in a single flash and, for all his subsequent study, he added nothing, save only the realisation of the particular significance of beacon hills as terminal points in the alignments."

The more academically inclined writers on these subjects have failed to attach any real importance to this feature of the phenomenon and indeed, have questioned the accuracy of the above account. The rational logic of academic minds stems from a different half of the brain from revelatory perception, so perhaps visionary experience does not come as easily to them as it does to a more spontaneously creative mind.

Visions tend to be subjective and often assume immense personal significance to those who experience them, sometimes changing the course of their lives completely By contrast, those who have never experienced anything remotely like it themselves may not see the relevance at all. Worse still, they may assume that it's an ominous symptom of impending mental instability, or at the very least, an excuse for ridicule. Anyone with even a passing interest in these subjects is frequently assumed to have joined the ranks of the cranks.

Despite these obvious drawbacks, Watkins is not alone in admitting to a visionary episode. A great many people openly acknowledge that their interest in ancient sacred sites and associated enigmas was initially aroused by some kind of spontaneous mystical experience.

In TWELVE TRIBE NATIONS AND THE SCIENCE OF ENCHANTING THE LANDSCAPE by John Michell and Christine Rhone, the authors mention Jean Richer, whose discoveries of aligned sites throughout Greece and the Mediterranean were prompted by a series of vivid and unusually memorable dreams.

Dr. James Swan, in his book SACRED PLACES recounts how an unexpected, dream-like experience in Dakota's Black Hills initiated a path of personal discovery that has made him something of an authority in this unusual field. Dr Swan has since reviewed more than one hundred case histories of people who have had extraordinary experiences at power spots. He states that the most common occurrences have included feelings of ecstasy or of unification with nature, interspecies communication, waking visions, profound dreams, the ability to seemingly influence the weather, feeling unusual energies and hearing either words, voices, music or songs.

His books, as well as those by Jamie Sams, Ed McGaa and many other Native American authors, contain many personal accounts of revelatory episodes, as well as a wealth of enlightening information on how Native Americans utilise the different qualities of special sites for a variety of spiritual activities, including Vision Quests, undertaken to provide guidance or inspiration for a new direction in a person's life.

David Icke and Shirley MacLaine, of course, have both included in their books personal accounts which support the premise that sacred sites act as a stimulus for unusual experiences of a spiritual or visionary nature. Even Paul Devereux, former editor of THE LEY HUNTER and one of the most down-to-earth writers on Earth Mysteries, admits in SYMBOLIC LANDSCAPES to falling on his knees at Avebury and asking the spirit of the place to reveal some of its enigmas. Much to his surprise, he was subsequently enlightened by an important insight into one of the mysteries of the local landscape.

The apparent ability of sacred sites to communicate indirectly or produce mystical or spiritual experiences in susceptible individuals is becoming extremely well documented and the manner in which the Earthstars' discovery evolved demonstrates yet another example of the phenomenon.

It was prompted by a series of psychic experiences, spontaneous dreams, and curious synchronicities, many of them at, or directly associated with, the specific places which form the initial mark points of the Earthstars' design. This was not something I consciously set out to discover. It was as if a series of clues had been deliberately arranged to arouse my curiosity and encourage me to make the discovery, piece by piece, like putting together a large and complex jigsaw puzzle.

Although there are certain techniques which can help improve an individual's receptivity to intuitive or psychic information, the revelatory process did not appear, at the time, to be one over which I appeared to have much control. The various pieces of the puzzle came together little by little over a long period. When I had assimilated one piece of information about a site or its related alignments, weeks or months might pass uneventfully before another experience or synchronicity would trigger some new facts about another site, or another level of connections, alignments and patterns.

Much of the revelation came during 1982 and 83, but once begun, it evolved into an on-going process which continues to this day and has become a very important and meaningful part of my life.

To me, Earthstars was not a discovery in the normal sense of the word. I have always regarded it more as a release of knowledge from the Earth itself.

I realise that, to many, all this may seem extremely far-fetched, if not totally unbelievable. In the context of the extraordinary phenomena associated with sacred sites, it is only to be expected.

It is a sad reflection of our times that visionary experience is currently regarded as something odd, something to be avoided, something that incurs the risk of you being labelled a loony.

People who see visions and hear voices in this day and age are not automatically assumed to have undergone a spiritually enlightening experience. More likely, they are assumed to be suffering from some form of mental illness, possibly schizophrenia.

However, far from losing my senses, my involvement with the hidden, spiritual dimensions of nature helped me to regain senses I never realised I possessed: senses that those of us who have grown up in a world dominated by the materialistic perspectives of business and commerce need to regain if human development is to pass beyond the chaotic, self-interested and self-destructive phase that passes for normal life these days.

These crucial gifts are; our sense of the sacred in nature and in ourselves: our sense of union with the natural world in which we live: our sense of ourselves as spiritual beings, as well as physical organisms: our intuitive senses in general.

In societies where there is a deeper understanding of the nature and purpose of sacred sites and personal visionary experience, as with the North American Indians, Australian Aborigines and many other indigenous cultures, these things are not deemed to be exclusive to saints, eccentrics and the mentally unstable. The importance of dreams and visions is understood as a perfectly normal part of everyday life and a natural consequence of living in a sacred landscape.

According to Dr. James Swan, in these cultures, it is those who do not have visions who are thought to have lost their senses. Absence of the visionary faculty is taken as an indication that they have lost contact with their creator. In this respect, a large percentage of the world's populace is one sense short of a full set and our heavily built-up urban landscapes only help alienate us all further from creation.

It is time to accept that the Earth (not to mention the rest of the universe) has a spiritual dimension and that we are as much a part of it as the physical world. We can inhabit it consciously, or unconsciously.

We have a lot to learn about the hidden connections we have to the planet and to the universe at large.

We can learn from cultures who have never lost their connections and who maintain a tradition that stretches back to the roots we must share with all the Earth's peoples. We can also learn directly from the Creator through the Earth's sacred sites as I believe our ancestors did.

There is much more to these places than meets the eye. Here in

London, particularly in the city, our ancient sacred sites are now surrounded by an alien culture.

Once, when churches and cathedrals, or for that matter, stone circles, were the most important places of the community, they dominated the skyline and landscape. Now many of our sacred places are overshadowed by ugly office blocks and other, soul-less developments, their origins and importance overlooked.

A materialistic society has developed around them, judging them on appearance alone and ignoring their spiritual dimensions.

Beneath the bricks and mortar, beneath the paving slabs and tarmac, these are still places of power, places of healing, places of the gods, places of vision.

What has happened to them, mirrors what has happened to us and to our society. We have become separated from our origins, our heritage and our natural environment.

This is a book about connections. But it goes way beyond simple connections between old churches. The most important connections to be made to our sacred places are our own.

Through them, we can re-connect to the hidden unity which links all life.

The secrets of the universe are not hidden.

On the contrary, they are on permanent display all around us.

To understand your place in the overall scheme of things,
you first have to understand the nature of the place
you are in.

You need to discover you are a part of it.

We are not separate.

People and planet are one.

Every atom, every molecule of your body belongs to the Earth
and will remain here long after your soul has departed for
other realms.

You are a spirit clad in Earth,
dust and water walking.

The mysteries of the Earth
are the mysteries of life itself.

Any investigation into the nature of the Earth's sacred sites
will, of necessity, evolve into a path of initiation
through the spiritual dimensions
we share with our planet.

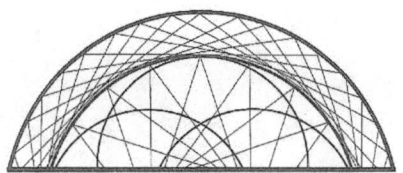

Chapter One
A Wake Up Call

**The Earth speaks to us. All you have to do
is find the right place,
to sit still and listen.**

Places of power have a way of calling you to them. Don't ask me how. The first time it happened to me was back in 1976. I woke up one Saturday morning with an overwhelming and inexplicable compulsion to go to Stonehenge. I'd never visited the place before and, up to that point, hadn't been particularly interested in it. I wasn't even sure exactly where it was. Nevertheless, the idea of going to see it had taken hold. I phoned a friend who lived nearby to see if she'd like to come along and as she had nothing better to do, we consulted a map and set off.

The old Vauxhall I drove at the time broke down before we'd even got out of London, but a little problem like that was no deterrent to an irresistible impulse. I called out the RAC and a couple of hours later, the old motor was rejuvenated with a new set of plug leads and carried us safely to the heart of Salisbury Plain.

It was a relatively uneventful visit, but for two things. In those days, you could wander freely amongst the stones. As I did so, the close proximity to them created two distinct impressions.

First, the stones had a tangible presence. Several of them felt as if they were somehow alive, or gave the slightly unnerving impression that someone, or something, might be inside the stone. One huge sarsen in particular felt as if it had a definite identity. For reasons totally inexplicable, it reminded me of Merlin whose magic, in legend at least, had created this monument.

At the same time, I acquired the strange notion that the stones themselves somehow held information about the purpose and past history of the place, information that could be accessed intuitively, by a kind of psychic archaeology.

Since then, I've learnt that this was hardly a novel idea. Native

Illustration 1: Stonehenge.

American Indian traditions regard stones as record keepers which can store place memories as well as other information. These days, of course, modern technology makes great use of information stored in quartz, but back in 1976, the idea struck me as extraordinary and quite intriguing. I had no idea of exactly how this information might be accessed, but I settled down on the grass and tried to relax into a receptive frame of mind to see what transpired.

Ostensibly, nothing happened. I saw no startling clairvoyant images, received no blinding flashes of inspiration. Yet, looking back, I was never quite the same person after that visit. For one thing, I had caught a minor dose of what John Michell describes as megalithomania; an insatiable fascination for the mysteries of stone circles and other ancient monuments. As afflictions go, it is relatively harmless, but quite incurable. I'd also acquired the notion that ancient sites stored their own history and other information as a kind of place memory which may be accessible to the intuitively sensitive. With the benefit of hindsight, I suspect that, in picking up this single idea, I had already acquired the one piece of information the site held for me at that time and, as Albert Einstein put it; **"Once expanded by a new idea, the human brain never regains its former dimension."**

Slowly but surely, my life and interests began to change. When I returned to London, any material connected with Leys, Earth energies, ancient sites and any related subjects was eagerly consumed. The seed had been planted and lay dormant, waiting only for the right time and place to burst into life.

In the meantime, I began to develop an awareness of my own spiritual and psychic dimensions. Two things seem to have acted as catalysts to this process. One was the Tarot. I'd been given my first pack of the cards as a birthday present in 1979 and when I tried them out on friends, I was surprised to find that the cards revealed a reasonable amount of accurate information. Intrigued by this success, I began studying the Tarot more seriously and found there is much more to it than simply a tool for fortune tellers.

The second big influence in this area was my father, who came to visit one day and instantly convinced me of the reality of life after death. At the time, he'd been dead for several years.

On many occasions, I had caught clear, but fleeting glimpses of him or sensed his presence around the house. After a while, I eventually realised that his frequent appearances meant that he had something important to communicate but I wasn't capable of picking up the detail of what he wished to pass on, only the sense of an urgent need to communicate something. A friend suggested I should arrange a sitting with a medium at The Spiritualist Association of Great Britain in Belgrave Square. It was the first time I had done anything like that and so had no idea what to expect. When I phoned to make the appointment, I was simply booked in with whoever happened to be available.

The medium, in this instance, turned out to be a softly spoken Scottish gentleman. He had never set eyes on me before in his life, but this didn't prevent him from describing my father with absolute accuracy when telling me that someone " in spirit" had accompanied me to the sitting in order to communicate from the other side. He then proceeded to listen to my father and relay what he had to say.

The degree of the detail was astonishing. It included how he had died, what type of car he drove, some of his interests, what our family dog was called, plus a whole lot of personal messages for my mother and other family members which included information only my father could have known. It was a life-changing experience. I came out of the sitting with eleven pages of notes and a firm belief in the reality of the spiritual realms.

I had always been interested in out of the ordinary subjects, like yoga, mysticism, psychic abilities, even UFO's, but this experience acted as a strong catalyst and I began attending the regular lectures held at the spirit-

-ualist Association in Belgrave Square and The College of Psychic Studies in South Kensington, where I later joined a psychic development class under the guidance of one of their resident mediums, Robin Winbow. I also began learning to use simple exercises in meditation and visualisation designed to develop or improve psychic ability, blissfully unaware that my own spiritual dimensions might extend much further into the world around me than I realised.

In the past, I'd had occasional psychic experiences, but now they began to become more frequent and whilst most would be irrelevant to this story, one in particular seemed to trigger a sequence of events which, in retrospect, might be considered to have sparked the Earthstars' discovery.

One of the very first methods of meditation I practised was to light a candle and, by taking deep, slow rhythmic breaths, relax and focus attention on the candle flame. I found the method in a book on yoga and for a consciousness-altering exercise, it sounds pretty simple. At first, it produced no obvious results, then on one particular evening, its effects escalated quite remarkably.

The glow around the candle flame grew brighter and changed colour until the candle was engulfed in a ball of rich orange with flashes of iridescent colours around it; a rich royal blue, or an emerald green with a golden centre. Occasionally, a violet tinge appeared at the edges. I had noticed similar effects on previous occasions and it was a fascinating phenomenon, quite relaxing to observe. This time, however, the experience took an unexpected turn.

The orange glow became distinctly gold, with an astonishing green and violet flare at the top. In the centre of the light, appeared the image of a woman, apparently wearing a long white robe or dress. She was holding her arms outstretched in a welcoming gesture, a startlingly recognisable pose often employed in statues of the Virgin Mary.

The image remained relatively small, maybe a few inches high at the most, but exceptionally clear. I remember thinking at the time that if I were a devout Catholic with no other religious influences, I might assume unquestioningly that this was indeed some kind of apparition of Mary, whose visitations seem to be frequently reported from all around the world. There was no verbal communication, but on an emotional level, she certainly made her presence felt. The figure had an aura of goodness, warmth and affection of incredible intensity. She literally radiated a feeling of love, peace and happiness. It felt like I was in the presence of a being of great beneficence.

As I watched transfixed, a ball of energy, rich royal blue in colour, gradually formed between her outstretched hands. It slowly grew into a sizeable sphere approximately five to seven inches in diameter, then floated lightly towards me. When the sphere reached me, it dissolved with a faint tingle against my solar plexus.

Overall, the experience felt energising and very uplifting. I wasn't capable of drawing any immediate conclusions about what I was seeing. It was an experience of deep emotional feelings, rather than rational thoughts.There were no words, no momentous messages. Just a sense that something very important had happened, even if I didn't quite understand what. I had no obvious explanation for it and no conscious understanding of who the white lady was.

As I said, I was aware of her startling similarity to classical depictions of the Virgin Mary. On the other hand, I had read Robert Graves' book THE WHITE GODDESS some years earlier and wondered if the apparition may have some connection to the mythic goddess who had been the inspiration for his title. She too represented a virgin, her purity reflected in the stainless white outfits that were her trademark.

I suspected the two may have a common source as a kind of spiritual energy which was able to manifest as a representation of any feminine deific figure and which was seen in different ways by people of different times and cultures.

I didn't mention the experience to many people, only one or two friends who were interested in spiritual and psychic matters themselves. None had experienced anything similar, although Beryl Bohea Raine, a medium I had met through The Spiritualist Association in Belgrave Square, did suggest that the ball of blue light probably represented spiritual energy, being transferred to me for healing and that the Lady may be a personal spirit guide. Certainly, I felt a strong affinity to the figure and had a feeling that her influence was entirely benevolent, almost as if she were some kind of guardian angel.

With the wisdom of hindsight, the episode reminds me of the sequence in Star Wars where Princess Leah's tiny holographic image inspires Luke Skywalker to set forth on his quest. Except that in the film, Luke was given some idea of what he was getting himself into and what to do next. At that stage, I didn't have a clue where this might be leading. Like the visitations of my father's spirit, I felt that the apparition of the lady in white had something important to communicate to me, but I had no idea what.

Although I repeated the candle meditation on a regular basis, she never materialised again in the same way Instead, a few weeks later, she appeared in a dream.

Dreams can be strange and complex things. This one wasn't. It was very simple which I suppose made it more vivid and easier to remember. Then again, it didn't feel like a dream. Dreams are often woolly and disjointed, full of things that don't make sense. This one had a curious sense of reality to it and that is what made it stick in the mind long after the average nocturnal fantasy would have faded.

In the dream, the white lady appeared as a radiant figure against a rich green background of bushes and trees, as if she was in a woodland setting. Her pose was much the same as in the candle flame vision, but this time, she spoke. Just one sentence;

"Seek me at Camelot."

Again, the experience carried an overwhelming sense of importance and despite its enigmatic aspects, I felt I should take the message literally.

Several locations vie for the title of Camelot, including Tintagel Castle in Cornwall, the hill fort of South Cadbury in Somerset, Caerleon in South Wales and Colchester in Essex, formerly Camulodunum in Roman times.

Intuitively, I knew that the message didn't relate to any of these places. I felt it was directing me towards a particular spot a little nearer home: an ancient moat on the Hertfordshire borders of North London that once bore the name Camelot but had attracted very little public attention over the centuries.

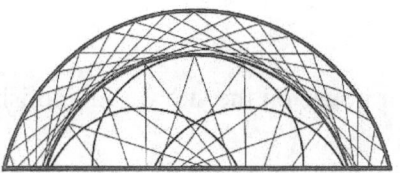

Chapter Two
London's Lost Camelot

On the crest of a hill, hidden in secluded woodland on the very fringe of North London, lies a small, moated isle. It's known locally as Camlet Moat, but over the centuries, the name has been abbreviated. It was originally called Camelot and, according to some sources, the name dates back to the 12th Century, when the Manor House of the Earl of Essex, Sir Geoffrey de Mandeville is thought to have been located there.

David Pam, a prominent local historian tells us in his book, THE HISTORY OF ENFIELD CHASE, that a survey of the area conducted between 1656 and 1658 states:

> **" The Manor and Chase of Enfield were anciently in the possession of Geoffrey de Mandeville in the reign of William the Conqueror, whose seat and habitation at that time, called Camelot, was situated on the chase near unto Potter's Lodge, the ruins whereof are yet remaining and being moated is to this day called Camelot Moat."**

A map from the same period confirms the name and shows that Camelot was not just the moat, but a huge area around it. The location of the moat itself is clearly identified as Camelot Hill. Surrounding it are North Camelot, East Camelot and West Camelot. The road passing the moat's north side is marked as Camelot Way and although parts of it have since been re-named Ferny Hill and Hadley Road, a long section running through Hadley Wood still preserves the original name as Camlet Way and leads directly from the Parish Church of St. Mary the Virgin, Monken Hadley.

Where Camlet Moat's name originated is something of a mystery. The above quotation suggests that it dates right back to the time of Sir Geoffrey de Mandeville's ownership which ended in 1144. However, the earliest reliable reference to the location as Camelot is in the fifteenth century when a fortified manor house on the site was demolished. David Pam's book states that;

> **" In May 1439, instructions were issued that the Manor of Camelot should be taken down, the materials sold and the money employed towards the repair of Hertford Castle."**

It is ironic that the most reliable records we have of Camelot's existence relate to its destruction, but at least it would be safe to assume that the name must have been attached to the Manor for many years prior to its demolition.

According to local historians, the most likely candidate for naming the place Camelot was not Sir Geoffrey de Mandeville, but one of his descendants, Humphrey de Bohun, the last owner of Camelot Manor. If this is correct the London's Camelot was destroyed at least a generation before Mallory's found fame with the publication of his epic Morte D'Arthur in 1470.

Alternatively, if the name Camelot really has been associated with this spot since Sir Geoffrey de Mandeville's day, it pre-dates Mallory by over two centuries and earlier Camelots by an equally impressive period.

According to Graham Phillips, author of THE SEARCH FOR THE GRAIL and KING ARTHUR - THE TRUE STORY, Chretien de Troyes was the first writer to use the name Camelot in any Arthurian saga. His works preceded Mallory's and began circulating in France between 1160 and 1180.

Since Geoffrey de Mandeville died in 1144 at the siege of Burwell Castle in Cambridgeshire, these would not have been the inspiration for his Camelot, unless there was some pre-existent tradition of which Chretien de Troyes and Geoffrey de Mandeville were both aware.

This is quite possible. It is well known that the authors of the various Arthurian tales, including Geoffrey of Monmouth whose HISTORY OF THE KINGS OF BRITAIN appeared around 1135, claimed to have developed their works from much earlier manuscripts which, in turn, may have evolved from a long-standing tradition, passed on by word of mouth or possibly by family line.

We should also consider the simple proposition that the place was already called Camelot (or Camlet) when Sir Geoffrey took up residence as Lord of the Manor. It is not beyond the realms of probability that some version of the name pre-dates Geoffrey de Mandeville.

After all, the site itself almost certainly does. The mere name of Camlet Way suggests an association to the same remote era as the Fosse Way or Icknield Way and although the name alone is a tenuous link, Roman artefacts have been found in the area of Camelot Moat to substantiate this suggestion. In the 1920's, Sir Phillip Sassoon, who owned the estate at that time, claimed to have unearthed Roman shoes and daggers there, whilst Barnet Museum's records show that a Mr C. Houston sent them four Roman coins dating from the fourth century, which had been discovered in the vicinity.

Sadly, all of these finds have since been lost, but if they were genuine, and there is no reason to suppose they were not, the evidence would suggest that the area could have been occupied for many centuries prior to de Mandeville, including the fifth and sixth centuries, the correct time frame to be contemporary with a historical King Arthur, if there ever was one.

The fact that it merited use as the location for the Earl of Essex's Manor House, in itself, suggests that this may have already been a site of considerable importance prior to the Norman conquest. Although not referred to specifically, Camlet Moat is thought to have been included in the properties and lands granted to Sir Geoffrey de Mandeville following the Norman conquest. Previously, in the reign of Edward the Confessor, these had included the Manors of Enfield and Edmonton and were in the hands of Ansgar "Staller to the King" and Sheriff of Middlesex. Camelot Moat may well have been Ansgar's manor and centre of operations before de Mandeville appeared on the scene.

As well as taking over Ansgar's property, de Mandeville also acquired the title Constable of the Tower of London and assumed responsibility for the defences surrounding London

This is by no means conclusive, but if Camelot Moat has been occupied since Roman times, it may have constituted an important part of London's outer defences.

It could therefore have played some strategic military role during Arthur's struggles against the Anglo-Saxons who had over-run much of London and the South East following the Roman withdrawal. Some historians maintain that the name Camelot derives from an ancient British god of war called Camulus, so the associations here could possibly have the same foundation.

Worth bearing in mind, in this context, is the suggestion of various Arthurian authorities that Camelot was a moveable campaign base that went wherever the action was. In that case there would have been many Camelots

Illustration 2: Reproduction of a 1658 Map showing that Camelot originally covered a far greater area than just the moat. Camelot Moat is shown on Camelot Hill, surrounded by East, West and North Camelot (published most recently in David Pam's book, THE HISTORY OF ENFIELD CHASE).

and this may well have been a local King or Chieftain's base given over to Arthur when he was in the area.

All this, though, is pure conjecture. There is, sadly, no hard evidence that gives any real clue to Camelot Moat's origins, or to the origin of its name. Nor are there any Arthurian legends associated with the site, as far as I know. Unless some conclusive archaeological evidence comes to light in the future, it is highly unlikely that we could ever prove an association with Arthur's Camelot.

Nevertheless, it was someone's Camelot and whatever else it may be, it is very much a place of mystery.

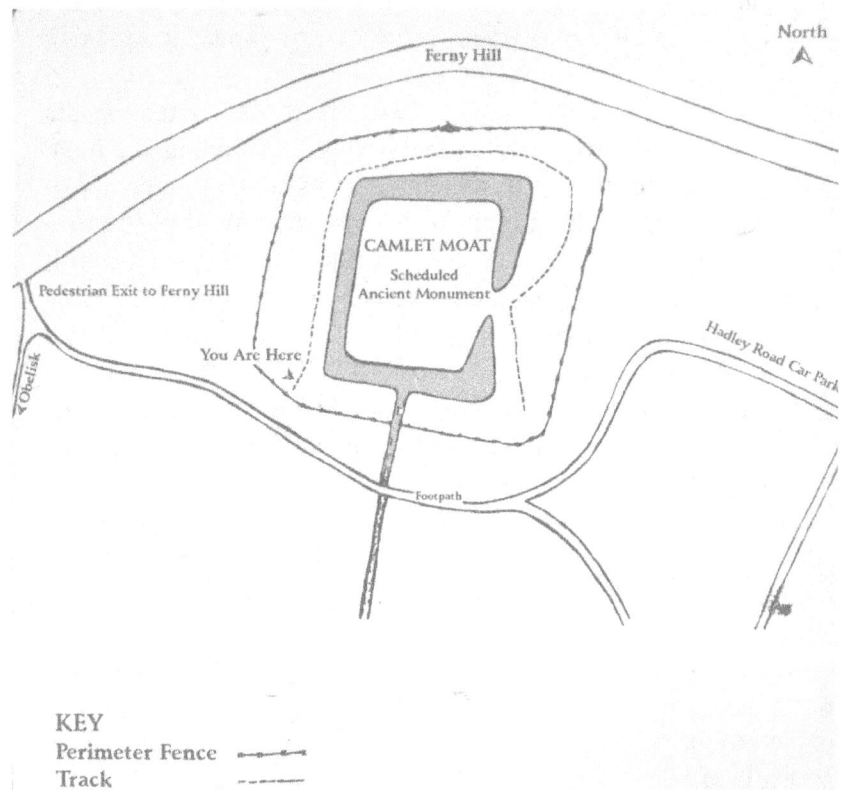

Illustration 3: Plan of Camlet Moat.

On my first visit to Camlet Moat, I knew nothing of its history. I'd simply seen the name on a local map and, following the White Lady dream, decided to take a look at it.

The moat is tucked away in woodland on the northern boundary of Trent Park, near Cockfosters in North London. The moat itself is not shown on some of the park maps, except as an area out of bounds to the public, presumably to discourage visitors who might damage an important archaeological site.

By car, Camlet Moat is best approached via Ferny Hill, formerly Camlet Way. Apart from the short stretch that curls around the moat, the road is reasonably straight.

When approaching the curved section, Camlet Moat is directly ahead of you. You can't actually see it from the road. It's obscured by trees. In fact, your eye is more likely to be drawn by a tall stone obelisk which stands incongruously on the edge of the field nearby, looking as if the woods behind might hide the overgrown ruins of a mis-placed Egyptian temple.

Nevertheless, it was not the obelisk that attracted my attention the first time I drove down that road. When the woods surrounding the moat came into view, I became aware of something unusual in the air above them. There was nothing visible in a physical sense, but I was aware of a kind of glow above the trees. Then, in a momentary clairvoyant flash, it became a huge and radiant image of the white lady, towering fifty or sixty feet high. The clarity of the image lasted only a split second as if it had simply been switched on and off.

The brevity of the image meant it was not quite as distinct as in the candle meditation or dream and the intensity of light and energy emanating from it gave the figure a similar quality to an over-exposed photograph but was clearly the same figure, the same pose, the same sense of recognition. I wish now that I had made a note of the date. Obviously, this kind of manifestation doesn't happen every day and I have never sensed it as vividly since.

As you can imagine, it was a weird experience. The image itself came and went in a split second, though the energy and presence could be still be felt as I drove on to the public parking area. The atmosphere felt positively charged with her presence.

I could still sense a kind of glow above the site ahead of me as I

Illustration 4: The N.E. corner of Camlet Moat.

trudged the short distance through the woods towards the moat. It didn't feel in any way unnerving. The atmosphere in the woods was surprisingly calm and peaceful.

I had never been to the place before and, unaware that one of the park's main paths passed close to the moat, I took what I thought was a direct route through the woods, stumbling through the undergrowth for an unnecessarily long time. At one point I remember feeling like the prince in Sleeping Beauty's fairy tale, looking for her lost castle hidden within an impenetrable thicket of bushes and brambles.

Eventually, I found the fence which at that time encircled Camlet Moat and, near a huge oak (now fallen), a broken section allowed easy access. The moat is about 15-20 feet at its widest, quite still, and carpeted in places with a thick layer of green algae. Some patches are just mud, reasonably solid-looking, until you step on it and squelch up to your ankles in smelly sludge. Looking over to the island, it seemed very overgrown, with plenty of mature trees and very dense undergrowth.

There was only one spot where it was possible to cross over and that was protected by a barrier of dense ferns almost six feet high, concealing trip wires of brambles and other hazards. There was no obvious path. I was clearly its first visitor for years, so I had to force my way uneasily through the vegetation before stepping across the narrow ridge of ground onto the island. It felt like I'd stepped into another dimension.

I have never experienced such a sudden and dramatic change of atmosphere. The hair on the back of my neck literally stood on end. Not because it was scary, although some people might have found it so. It was simply the intensity of the energy at the place. The air almost throbbed with it. Despite this, I did my best to look around. I noticed several roughly excavated trenches from an archaeological dig. Apart from that, the isle didn't look much different from the woods surrounding it.

On a non-physical level, it was extremely disorientating to the point where I felt I couldn't remain there for more than a few minutes. I had never experienced anything like it before. I had to leave. Even then, I had trouble finding my way back to the crossing point. As you can imagine, it didn't even cross my mind to attempt any deliberate psychic work at that moment.

Over the course of the following months, though, I visited Camlet Moat regularly, researched its local history and began a series of curious encounters with the white lady, who seemed particularly attached to the place.

THE LADY OF THE WELL

As Camlet Moat had come to my attention by psychic means, my attempts to glean further information about the place employed similar methods: frequent visits, meditation at the site and psychometry using stones picked up around the moat. Sometimes I saw quite clear clairvoyant images. More often than not, I wouldn't get anything particularly relevant. Quite frankly, the process of what was revealed didn't seem to be altogether under my control.

The white lady apparitions could be sensed at various locations on the isle, but seemed to be more strongly focused on the North East corner near the remains of a well. This was apparently once a construction with impressive stonework.

Today, it is hardly recognisable as a well, just a deep muddy hole. The reason is sad, but predictable. Legend had it there was treasure to be found at the bottom of the well. The treasure hunters here were the families who had previously owned Trent Park; the Bevans and later, Sir Phillip Sassoon, who carried out a second amateur archaeological dig in 1923. If any treasure was found, it went unrecorded.

The reported finds included oak beams which formed the basis of a drawbridge, Roman shoes and daggers, as well as mosaic tiles depicting a knight on horseback.

Whatever structure had been on the site must have been substantial. The Enfield Weekly Herald reported that Sir Phillip Sassoon's workers had unearthed the foundations of;

> **" a massive wall two feet below the surface and going down to a depth of eight feet. "**

> **" The length of the wall excavated at that time was around forty feet and, in some parts, five and a half feet thick."**

One section of the wall was described as being constructed of **"Huge pieces of stone"** while another part was of **"flints as big as a man's head."** This wall still lies deep in the ground. Some of it was still visible at the bottom of the isle's excavated trenches before English Heritage re-filled them in 1999. The current whereabouts of any other finds are, unfortunately, as much a mystery as the treasure itself. What really surprised me was that at Camlet Moat, the white lady apparitions occasionally looked remarkably similar to popular conceptions of Gwen-

Illustration 5: Two views of the remains of Camlet Moat's well.

Illustration 6:
An image of the Virgin Mary in the same pose as the "white lady" apparitions seen by the author.

-evere. I shouldn't have been surprised. The name Gwenevere translates from its original Welsh as **"White Spirit"** or more literally **"white breath of life"** and in the Arthurian mystery tradition, she is a Christianised version of the White Goddess. In the apparitions, her dress was long and white, loosely tied around the waist with a cord. Usually she'd be wearing a simple head-covering like those seen on statues of the Virgin. Occasionally at the moat, she had appeared in a pointed hat, hung with a veil.

The curious thing about the hat was that when viewed from the side, it appeared to have a single, conical point, but from other angles it was clearly a two-pronged affair with a curve between the points, reminiscent of a crescent moon. Although the conical hat gave her a Guenevere-like appearance, the double pointed version lent her a different, but equally recognisable identity, to me at least. I thought she bore a remarkable similarity to the image depicted on the second card

Illustration 7: Card No II of the Tarot Major Arcana, the High Priestess, the Keeper of the Mysteries.

of the Tarot Major Arcana, the High Priestess. The High Priestess figure is dressed almost identically to the white lady, with a crescent moon tiara reminiscent of the hat's twin peaks. It is generally taken to represent the feminine principle of the divinity, as an image of the Egyptian goddess Isis, or Sophia, goddess of wisdom, although it is sometimes referred to as The Papess, Pope Joan.

In Tarot traditions, the High Priestess is the Keeper of the Mysteries and guide to hidden secrets. She governs inspiration, intuition and messages contained in dreams, all the kind of things, in fact, I'd had some puzzling first hand experience of in recent months. In earlier cultures, she has been known by many names and has associations with Astarte, Isis, the Roman Diana, the Greek Artemis, the Welsh Cerridwen and Brigit or Bride, the bright one.

I am not particularly clairaudient, but on one occasion I did sense a name with the apparition; Isis Maria, a curious combination of the latinised divine mother and her earlier Egyptian counterpart who was mother of the sun-god Horus.

I believe that in times past, visions such as this were one of the factors that helped identify certain places as sacred. I assumed the white lady was a genius loci (spirit of place); literally a manifestation of the energy of the site and possibly a goddess figure worshipped at a shrine here thousands of years ago.

In this context, the presence of a well on the isle may be more than coincidental. Many holy wells were sacred to the White Goddess. St Bride's church in Fleet Street, for instance, was built upon the site of a well sacred to Brigit or Bride as far back as the sixth century. (The well is still there, but not generally open to the public.) Bridewell Penitentiary takes its name from the same source. The previous building on the site was a nunnery.

As a result of these (and other) clairvoyant visions, I became convinced that, at some time in its distant past, Camelot Moat had been a sanctuary of the ancient British White Goddess in one of her forms, probably Bride or Brigit, and that the well had been an oracular shrine as well as a place of healing and initiation. One vision in particular showed the site without its moat, but as a stone circle with a particularly huge megalith in the NE corner roughly where the well is situated.

The vision reminded me that "huge stones" had been found built into the foundations of the wall unearthed in Sir Phillip Sassoon's archaeological dig. I wondered if these were remains of a megalithic monument here. I wondered too, if the moat had been created later as a symbolic sacred isle, where one literally crossed over to the "otherworld" of the spirit realms, or of the gods.

The crossing of water to the land of the spirits or of the dead features strongly in many mythologies and there are several well known sacred isles around the British Coastline; Holy Island, Iona, Bardsey, St. Michael's Mount and the former isle of Glastonbury, to name but a few. In his book EARTH HARMONY, Nigel Pennick mentions this subject and states;

" It appears that people believed magical powers were contained or enhanced by enclosure by water. "

Moated sites like this one are often classified as Saxon homestead moats and the watercourse generally presumed to be a defensive measure, although they were actually of little use defensively, their size and structure

Illustration 8:
The Parish Church of St. Mary The Virgin, Monken Hadley.

being totally inadequate for any kind of protection against marauders. There is now speculation that their purpose is more likely to have been connected to the spiritual beliefs and practices of the time. Nigel Pennick concludes that the most likely explanation of their function was probably related to some kind of magical protection, in the manner of the shrines of India and Indo-China.

In the context of the pagan beliefs prevalent in Saxon times, this makes far more sense and it indicates this place could have been a ritual site, rather than simply a homestead. I felt very strongly that this had been a sacred place to our most distant ancestors. In one vision, I had seen white-robed figures following a processional route along Camlet Way to a sanctuary and holy well at what is now Camlet Moat. Their starting point had been at another shrine and holy well in Monken Hadley where Camlet Way begins and close to where the Parish Church of St. Mary the Virgin, Monken Hadley now stands. Prior to the building of St. Mary's church, there had been a Hermitage in the vicinity. This had passed into the ownership of the Benedictines of Waldron Abbey in 1144 when it had formed part of Sir Geoffrey de Mandeville's property. So ostensibly the only recorded link between Monken Hadley and Camlet Moat is that de Mandeville owned both and the fact that they are connected physically by Camlet Way. Centuries old maps show it leading directly from Monken Hadley church to the Moat.

I felt that Camlet Way had been the processional route I had seen in the vision and that it pre-dated Monken Hadley's monks by several centuries. I felt too that Monken Hadley's original religious community had been an order of women who had been keepers of an eternal flame that burned in a special sanctuary atop a nearby hill or mound.

Old maps actually show several mounds in the vicinity of Monken Hadley church but none remain. A possible reminder of the eternal flame does though. Monken Hadley's church is one of the few in the country with a cresset (beacon burner) atop its tower.

It owes its existence to the fact that, prior to the establishment of the church, this was the site of a beacon, supposedly lit to guide travellers through the forest, which in those days, covered most of the area. However, there is another remote possibility. The devotees of Bride or Brigit maintained a perpetual flame in her honour at her sanctuary in Kildare Eire. Monken Hadley's beacon may have originated from a similar tradition, as the eternal flame of Brig, Brigit, Bride, in her role as the lady of the local shrine.

As I mentioned previously, Bride also has an associated with healing wells and there may even have been a holy well dedicated to her at Monken Hadley. Flooding in the church crypt earlier this century resulted in the waters of a long forgotten well or spring being drained and diverted.

With the coming of Christianity, many sacred shrines were re-dedicated to the Christian equivalent of whatever local deity had been revered there, so it is understandable that places associated with the white goddess Bride may have been re-dedicated to the Virgin Mary.

My feeling is that Camlet Moat's well and Monken Hadley church both began as such shrines and that they are linked by more than just the remains of Camlet Way and the memory of Sir Geoffrey de Mandeville.

ARTHURIAN MYSTERIES

Apparitions of Gwenevere are not the only Arthurian Mysteries associated with Camelot Moat.

In Gareth Knight's book, THE SECRET TRADITION IN ARTHURIAN LEGEND, there's a map drawn by the poet and mystic Charles Williams. It's a very rough outline of Europe superimposed with the outline of a goddess figure called Brizen, coincidentally, another name for Brigit or the White Goddess.

Strangely, the map has very few places clearly marked on it: one of them is Camelot.

Even more strangely, Charles Williams had not placed Camelot where most people would; at South Cadbury in Somerset or at Tintagel in Cornwall, Caerleon in South Wales, or even Camulodumun (The Roman name for Colchester) in Essex.

He had marked it in the general area of North London, not a million miles from where Camlet Moat stands.

In the hope of finding an explanation for this, I wrote to Gareth Knight and asked whether Charles Williams might have been aware of Camlet Moat. He didn't think so and replied that it was more likely Charles Williams regarded London as Camelot. Why Williams should think this way is hard to imagine, though it's obvious from his poetry that he did. There is a line in Williams' poem, TALIESIN IN THE SCHOOL OF POETS, which confirms Gareth Knight's opinion;

**"Through Camelot, which is London in Logres,
by Paul and Arthur's door."**

The idea that London is somehow Camelot seems to stem from Williams' esoteric interests. He was a member of The Hermetic Order of the Golden Dawn, an esoteric society derived from Rosicrucianism and the teachings of Hermes Trismegistus which counted W.B.Yeats and the notorious Aleister Crowley amongst its members.

I don't pretend to understand what the order's aims or practices were. However, it is known that some esoteric organisations believe the Arthurian Legends conceal a mystery tradition of immense antiquity, associated with a pre-Christian goddess cult.

This was, in fact, the entire basis of Gareth Knight's book. I had bought it specifically to find out what esoteric associations Gwenevere and Camelot might have. I gradually came to the conclusion that Camlet Moat itself could be in some way implicated in the mysteries of this tradition and that Charles Williams' map is a definite clue to this. It shows the spirit of the land as a female figure (a goddess) super-imposed over the physical landscape of Europe.

If the goddess images I had encountered at St. Mary's in East Barnet and at Camlet Moat are thought of as somehow an aspect of the spirit of the land, Williams' map is an artistic expression of exactly the same notion and the identical goddess. I believe that's too much of a coincidence to be a coincidence.

It certainly raises a lot of questions.

Was Williams aware of this particular Camelot?

Was its location known to other members of the esoteric society to which he belonged?

Did Camlet Moat play an important role in this secret tradition?

We may never know for sure, but interestingly, two of Camlet Moat's owners may also have had an interest in the esoteric.

The first is Sir Geoffrey de Mandeville who is known to have had curious links to the Knights Templar. They in turn have been associated with various mystical practices including the worship of Sophia and are actually portrayed in some of the later Arthurian Tales as the Grail Knights.

Secondly, Sir Phillip Sassoon, the last private owner of Trent Park (where the moat is located), may also have had discreet mystical inclinations.

In his sitting room, he had depictions of classical gods on his walls. They represent three of the four alchemical elements, Earth, Air, Fire and Water. Moreover, a contemporary of Sassoon is quoted in THE HISTORY OF TRENT PARK as saying;

" Whatever gods he worshipped, they were not our gods. "

Some of these themes are expanded upon in a later chapter.

Suffice to say that they could fill an entire book and will, once I have finished this one (See London's Camelot and the Secrets of the Grail).

There is much more to Camlet Moat than meets the eye.

At the time, trying to unravel what exactly was behind its mysteries provided a useful stimulus for my curiosity and kept me firmly on a path which was to lead towards other mysteries, of even greater significance.

London   City of Revelation

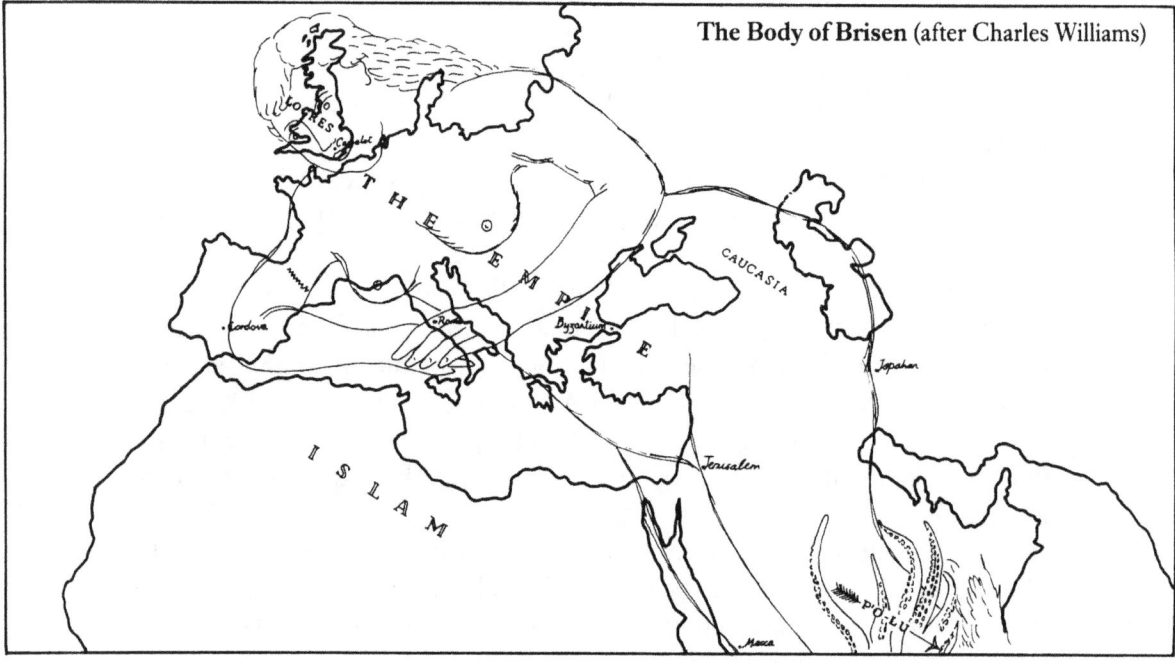

Illustration 9: Charles Williams's Map depicting Brisen, a namesake of Brigit or Bride, the White Goddess, as the spirit of the land and showing Camelot in the London area. Reproduced from Gareth Knight's book, THE SECRET TRADITION IN ARTHURIAN LEGEND.

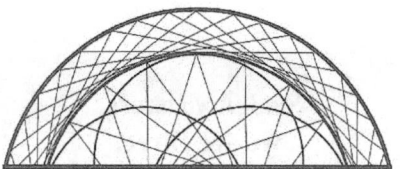

Chapter Three
The Dream

In the beginning, I had no inkling that Camelot Moat would be the starting point for a lengthy, life-transforming mystical quest, appropriate though the name is for such ventures. Nor did I realise that it was linked in intricate and intriguing ways to other ancient sites in the area.

The first clue to this aspect of the Earthstars discovery came in a vivid dream. It began with an image of the White Godddess. She looked much as she had in the original candle meditation but, as with the vision above Camlet Moat, her image towered high over a specific site; in this case, a small, hilltop church, almost hidden amongst the trees that filled its churchyard. I recognised the place immediately. I used to drive past it every day on my way to and from work. It was the old Parish Church of St Mary The Virgin in East Barnet.

The dream's focus of attention switched to the church tower and with the change of perspective came a clear and simple message;

" **This place is a spiritual power station.**"

Suddenly, a fountain of water began to gush from the top of the tower and cascade down its sides. As the flow increased, the tower become engulfed beneath a cascade which swirled through the churchyard and down the hillside, until the whole area was flooded under several inches of water. A second message accompanied these images, again in simple, memorable words;

" **These are the waters of Aquarius,
sent to wash over the land, to cleanse,
purify and heal the Earth.**"

Finally, the vision of the White Lady became three women dancing in a circle, then spinning together so that all three blurred into one.

**Seek my sisters,
we are three.
The lady is a trinity.**

Unlike the average dream, this one didn't fade on waking. Its images and three simple messages remained vivid and clear in my mind, so that now, years after the event, I can still recall them distinctly.

Its meaning seemed reasonably clear, too.

The notion that the Earth, at least in terms of the precession of the Equinoxes, was entering the age of Aquarius had been popular since the sixties and Aquarius is the water bearer. Obviously, the symbolism of the dream related to this. Its implication was that this site acted as a focus for some natural force within the landscape which could be represented symbolically as the "waters of life" and which had the power to refresh, renew and regenerate the land.

The average person might be forgiven for regarding such notions as the product of an over active imagination. However, subsequent developments did more to confirm than dispel them.

ST MARY THE VIRGIN, EAST BARNET

The Parish Church of St. Mary The Virgin, overlooking Oak Hill East Barnet is unquestionably the oldest church for miles around. It was founded around 1080 by the Benedictine monks of St Alban's Abbey.

I have always wondered why they should have chosen this particular spot for the first Christian church in the whole area. Topographically, the crest of the hill in High Barnet where St. John's Church now stands would have been a more obviously commanding site for the area's first church. But surprisingly, one wasn't built there until the 13th Century and even then, only as a chapel of ease to St. Mary's in East Barnet. Obviously, St. Mary's location was deemed to be far more important. What made it so special ? Had it been an important pre-Christian shrine ? Did the name Oak Hill, for instance, give a clue to a druid grove or venerated tree on the site ? This may be a possibility. Local folklore does actually include tales of at least one person having visions associated with an oracular oak on the hill.

More interestingly, one site alignment links St. John's in High

Barnet and St. Mary's in East Barnet to an ancient venerated oak in nearby Southgate.

On the O.S. map, this alignment appears to pass near to Christ Church in Waterfall Road, near Southgate Green. What the map doesn't show is that in a small, secluded walled garden to the west of the church lie the remains of the older Minchenden Chapel and a vast, impressive tree known as The Minchenden Oak. If the oak and the chapel occupy an earlier druidic place of ritual, it is hardly surprising that the alignment goes through them, not the nearby church which dates only from the last century. The alignment linking the three sites raises the possibility of all three being related spots of pre-Christian druidic importance.

Another clue which points to St Mary's as a location of special interest is the fact that an alignment through two of North London's sacred hills, Primrose Hill and Parliament Hill, points directly to it.

If a line is drawn from the summit of Primrose Hill to Parliament Hill, it can be extended directly to St. Mary's on Church Hill, East Barnet. The alignment is due North, which may have held some added significant for our pagan forebears. Some of their belief systems assign

Illustration 10: The Parish Church of St. Mary The Virgin, East Barnet. A spiritual power station ?

correspondences to each of the four quarters. The North is often associated with winter, with Earth as one of the four alchemical elements and with the place of the mother (possibly because in winter, much life withdraws into the body of mother Earth).The latter point was to prove extremely relevant to subsequent psychic experiences at the site.

Following the dream, I became a frequent visitor to St. Mary's in the hope of gathering further intuitive impressions and acquiring a clearer understanding of what might be behind these apparitions and experiences. St.Mary's picturesque little churchyard has a powerful presence and despite the gravestones, it's full of life, providing homes for large squirrel and bird populations. It also provides some peaceful spots to sit, relax and soak up the atmosphere of the place.

At St. Mary's, clairvoyant images of the goddess figure could be sensed, but in a slightly different form, looking more like the Empress card of the Tarot, with dark hair, a crown or tiara of stars and a dress with red floral emblems on it, probably roses. Here, she appears seated on a throne, a seat of power, rather like the images of Our Lady of Walsingham. In Tarot symbolism, the image represents the Great Mother, Mother Earth or the mother archetype in general. Local history records actually refer to St. Mary's as the Mother church in its relationship to the later St John's in High Barnet.

I felt that St. Mary's had been built upon a site that had, in pre-Christian times, been used as a place of worship of the Great Mother later re-dedicated to her Christian counterpart the Divine Mother, Mary. Although many modern day Christians may regard this as a heretical suggestion, it's not exactly a revolutionary idea. There is, in fact, a lot of evidence to suggest that the earliest churches in this country occupy sacred sites that had been the focus of religious activities for centuries before the arrival of Christianity, a large number of them hilltop sites like St. Mary's.

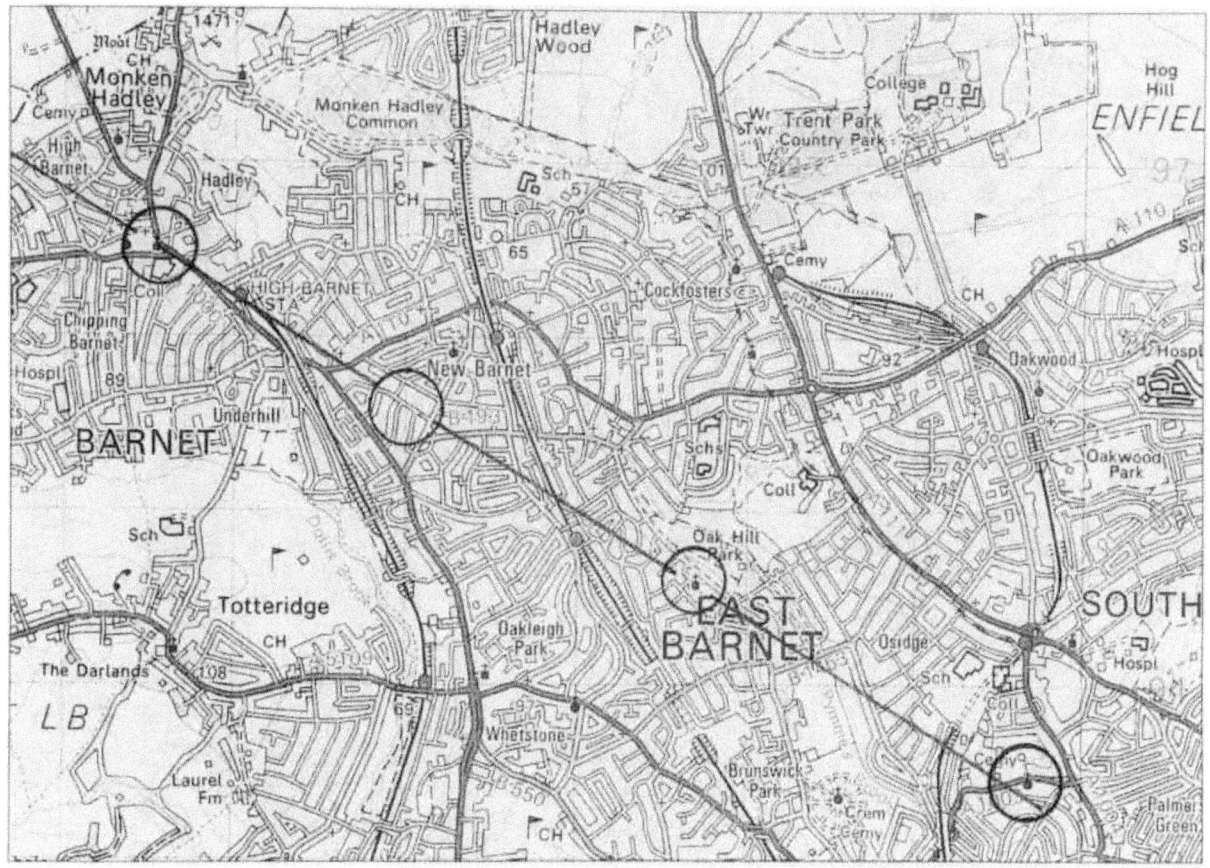

Illustration 11:

The alignment from St. John's in High Barnet through St. Mary's, East Barnet extends to the site of the Minchenden Oak in Southgate.

Illustration 12 (P38):

Primrose Hill and Parliament Hill align directly due North to St. Mary's on top of Church Hill in East Barnet.

O.S. map © Crown Copyright. All rights reserved. License No. 100029558

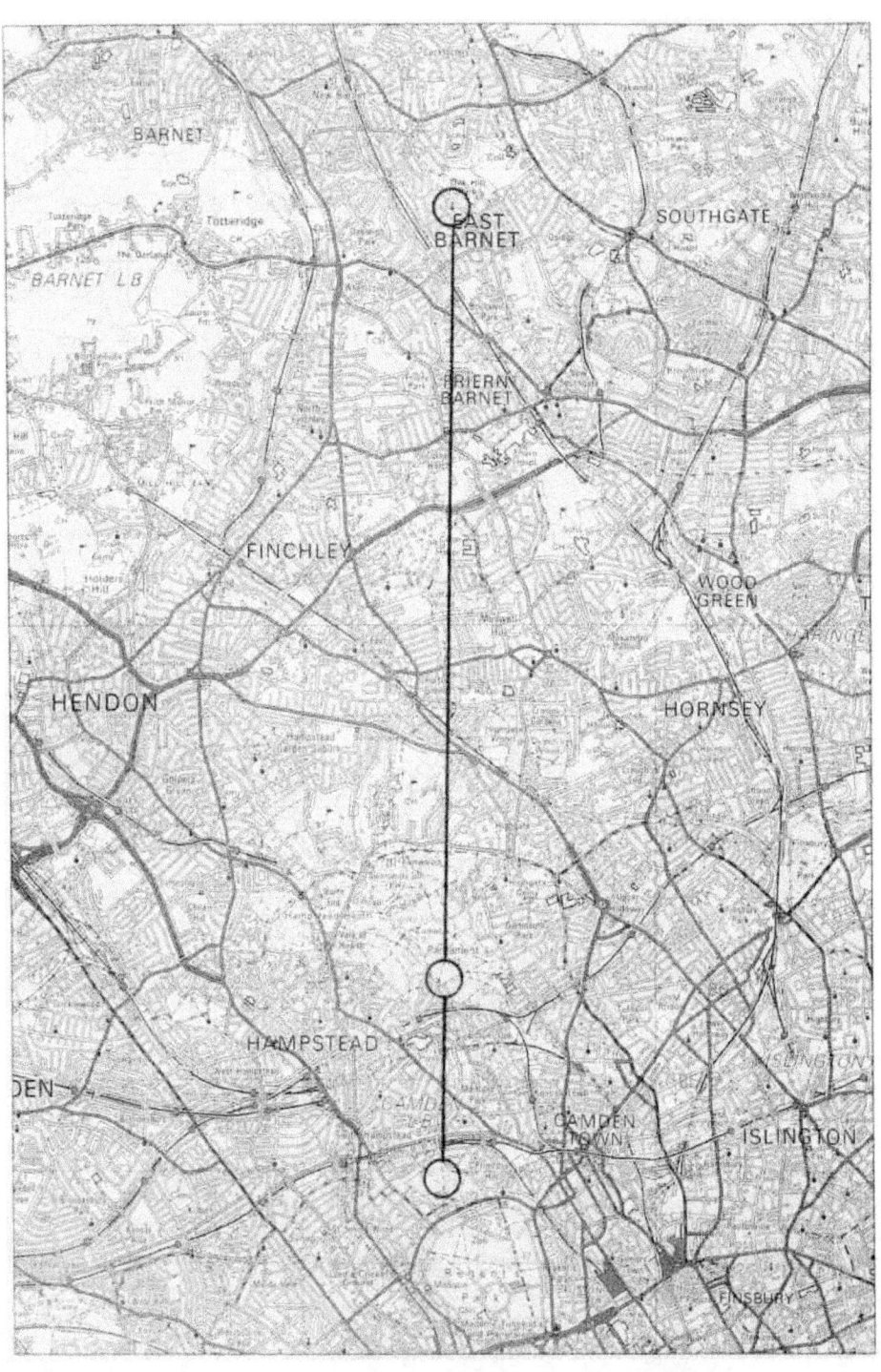

O.S. map © Crown Copyright. All rights reserved. License No. 100029558.

Illustration 13: The Empress of the Tarot Major Arcana. Clairvoyant imagery at St. Mary's resembled this figure which psychologically represents the mother archetype and bears some similarity to Marian images like those of Our Lady of Walsingham.

A SPIRITUAL POWER STATION ?

In many cultures, past and present, sacred sites are accepted without question as places of spiritual power. They were, and still are, places of the gods, places of vision, places of healing. They are locations where the gods may be seen, the power of God experienced or the spirit world contacted; where divine guidance may be obtained either directly or indirectly through oracles; places where the power of the gods could be felt. Some indeed were places where the holy water from a well or a spring was believed to have special curative properties. From this perspective, neither of the first two messages of the dream are telling us anything particularly new, except that these things may not have been previously associated with the little church of St. Mary The Virgin in East Barnet, at least not in recent times.

Churches of this antiquity are invariably built upon pre-Christian places of worship, so it is probable that our ancestors recognised this spot as a place of power and deemed it holy ground long before the Benedictines arrived and built a small church here.

After my experiences, I wondered if this was a known place of vision and healing to our ancestors. Local history does record one instance of a person deliberately sitting beneath an oak on the hill in the hope of receiving inspiration or a vision and if I could see the spirit of place as an Earth Mother /Great Mother archetype, presumably others before me who lived much closer to nature could have done so too, which brings us to the third message of the dream; the three sisters.

Triple goddesses are a recurring theme in mythology, particularly as representations of the cycles of nature. In connection with the Earth Spirit, who has generally been worshipped in the past as a Mother figure, the triple goddess represents Mother Earth's three active seasons: Spring, Summer and Autumn.

As a moon goddess, she represents the three phases of the lunar cycle; new moon, full moon and waning moon. Lunar cycles being directly linked to women's cycles of fertility, the triple goddess also represents the three phases of womanhood; firstly, virgin or bride; second, mother and universal provider and finally, the old wise woman. The Christian equivalents are; Mary the Virgin, Mary the Mother of Jesus and Mary Magdalene.

Until these experiences began, I didn't give mythic characters and ancient goddesses a second thought. Like most people, I suppose I assumed that they didn't really exist and were the imaginative invention of our ancestors. The usual explanation for the classical pagan pantheons is that they symbolise forces at work within nature which were personified with human characteristics to enable their functions to be grasped more clearly.

As my experiences with the goddess apparitions evolved, I developed a more literal understanding of this concept. I had no doubt that the goddess figures I regularly witnessed were manifestations of spirits of place, but in a wider sense were personified aspects of the life force of the planet, which Robert Fludd called the World Soul but which had been known to other mystics throughout the ages as the Earth Spirit, Mother Nature or Mother Earth.

I believe the forces or energies represented by our mythological figures, though intangible in the extreme, can be seen as appropriate personifications when encountered by someone like myself, with some degree of clairvoyant abilities. Moreover, these abilities may have been far more common in our ancestors and the special atmosphere of certain locations may enhance them considerably. Many psychic friends who I have introduced to these places have also seen comparable manifestations. It is, I admit, a difficult idea to grasp if you have not witnessed anything similar personally, but don't dismiss it out of hand. In EARTH HARMONY Nigel Pennick speaks of the noted archaeologist T.C. Lethbridge who was an important figure in the investigation of dowseable Earth Energies. I was enormously reassured to find that Lethbridge's conclusions were remarkable similar to my own. He believed that the concentration or generation of energies at certain places on our planet's surface could, under certain conditions, be perceived as **"gods, saints, nature spirits, elementals, fairies, ghosts, boggarts or demons."**

We seem to be dealing here with a maleable energy which interacts with the subconscious perception of the viewer to draw on and reflect a recognisable image.

Nigel Pennick adds that in recognition of this "traditional perception" Lethbridge called his dowsed energy fields after the nature spirits of the ancient pagan Greeks. Possibly what I was sensing as Mary, Guenevere, Isis or Tarot images, he would probably have called Athena, Artemis or Aphrodite.

Australian Aborigines would understand this general concept perfectly in terms of the dreamtime, their own visionary landscape, with its places of power and ethereal populace of gods, ancestors, power animals and spirits of place. So would North America Indians who traditionally engage in "vision quests" at appropriate sacred locations and regard the Earth generally as a spiritual mother who is able to manifest herself personally to them.

In the Lakota Sioux language, Mother Earth is Ina Maka. In Peru, she is Pacha Mama. The concept is universal, but the psycho-spiritual relationship is, of necessity, a personal one, as it would be with any aspect of the creator. That's why this book takes a spiritual approach to these mysteries. It has been a spiritual quest. Not a scientific investigation. This aspect of it should not really surprise anyone. These phenomena stem directly from the Earth's sacred sites: places where we are supposed to go to commune with our gods. It is only to be expected. Any investigation into the magical qualities of these spots is likely to lead to personal experiences involving our hidden spiritual dimensions.

I have no doubt whatsoever that including this type of subjective psychic material in the book will make me an easy target for those who want to ridicule or poke fun. Nevertheless, I felt it was absolutely necessary for a full understanding of how the Earthstars discovery actually came about. The visions, dreams and psycho-spiritual elements were an intrinsic part of it. Without them, there might not have been any discovery at all.

THE BARNET TRIANGLE.

Mention of a third "sister" in the dream suggested a third important location in addition to Camlet Moat and St. Mary's, East Barnet where apparitions of the first two "sisters" had materialised.

I was already aware of another site with connections to Camlet Moat; the old church at Monken Hadley Coincidentally, it was another St. Mary's. The place had been physically linked to Camlet Moat for centuries by Camlet Way which once ran directly from Monken Hadley church to the moat although the road undergoes two name changes en route these days.

Historically, the two also had a common connection through Sir Geoffrey de Mandeville, who owned both sites in the mid twelfth century. Back in those times, Geoffrey de Mandeville is known to have been associated with the Hermitage at Hadley, a monastic establishment whose chapel is thought to have evolved into the local Parish Church.

Curiously, all three sites were linked in various ways to the Benedictines. St. Mary's, East Barnet was founded by those from St. Alban's Abbey and both St. Mary's, Monken Hadley and Camlet Moat were loosely associated with Waldron Abbey's Benedictines through Sir Geoffrey de Mandeville who had been the founder of the Abbey and had passed ownership of the Hermitage at Hadley into their hands. Waldron Abbey has long since ceased to exist, but Audley End House at Saffron Waldron now stands on the site.

The first St. Mary's church at Monken Hadley dates from the fifteenth century. A coincidence of minor personal significance intrigued me. The current church was re-built during the last century and had been re-designed by a namesake, the eminent Victorian Architect, G.E. Street, who was also responsible for London's Law Courts.

A far stranger connection linked the three sites upon the landscape, though. After visiting all three in one afternoon and puzzling over what they might have in common, I sat down to plot any possible ley alignments through them on a map. A ley, for those who don't know, is supposedly a perfectly straight alignment of four or more mark points like standing stones, tumuli, pre-reformation churches and other ancient sites, sometimes stretching for miles across the countryside. A Herefordshire businessman and local magistrate, Alfred Watkins, coined the term back in the early 1920's and they have been a constant cause of controversy ever since.

To my surprise, the exercise in ley hunting didn't reveal anything particularly noteworthy as far as simple straight alignments go. It turned up something far more unusual. I found that Camlet Moat, St. Mary's East Barnet and St. Mary's, Monken Hadley formed an almost perfect equilateral triangle on the landscape.

As soon as I saw it, a shock ran through me. I had a strange sense of recognition, as if I'd known subconsciously what I had been looking for. I knew immediately that it was the first piece of a much larger puzzle. Later developments were to prove my intuition right, revealing the Barnet Triangle's role as the key to a geometric pattern of enormous complexity.

I was so surprised by this triangle that, at first, I overlooked the obvious symbolism. I had been informed that the goddess was a trinity. An equilateral triangle traditionally symbolises the Holy Trinity.

The Christian Trinity, of course, is Father, Son and Holy Ghost rather than a trinity of goddesses, but I couldn't help wondering whether the same triangle had symbolised the triple aspects of the Mother Goddess in previous cultures where she was the principal divinity.

BARNET TRIANGLE DIMENSIONS

Camlet Moat to St. Mary's, East Barnet; 2.375 miles approx.

Monken Hadley St. Mary's to Camlet Moat; 2.4 miles approx. St.

Mary's, East Barnet to St. Mary's, Monken Hadley; 2.425 miles.

Perimeter;
About 7.2 miles = 67.6 furlongs = 12,672 yds = 38,016 ft
= 456,192 ins = 2,304 poles (5.5yds) = 21,806.5 Royal Cubits

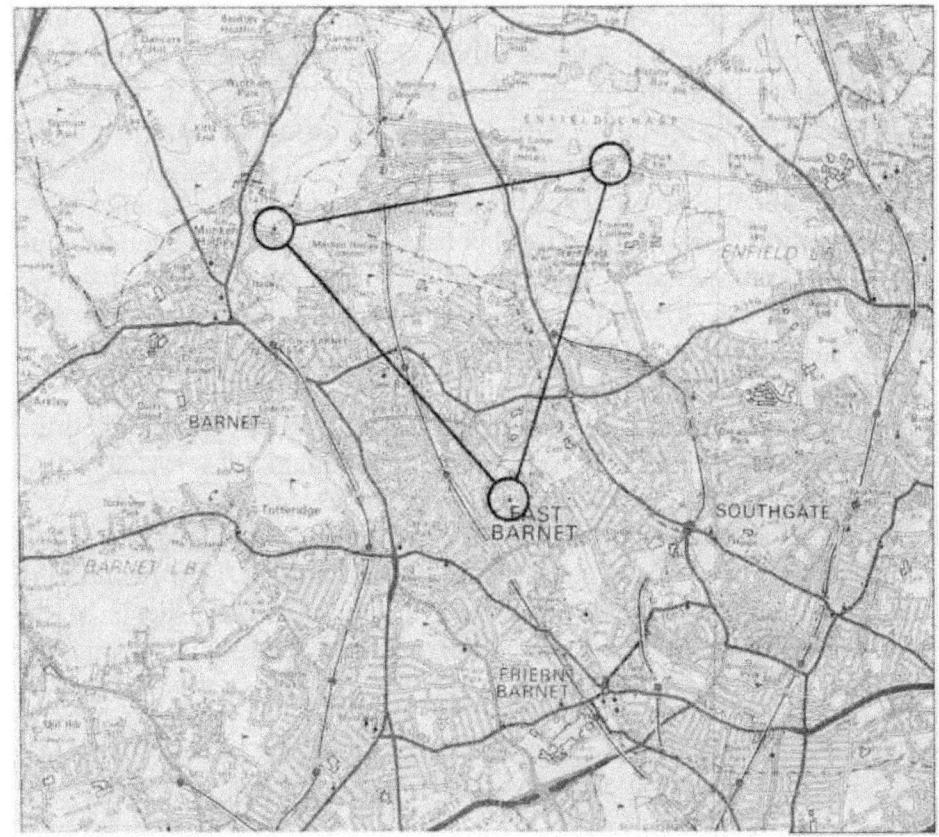

Illustration 14:
The Barnet Triangle. An almost perfect equilateral triangle defined by three of the oldest sites in the area

O.S. map © Crown Copyright. All rights reserved. License No. 100029558.

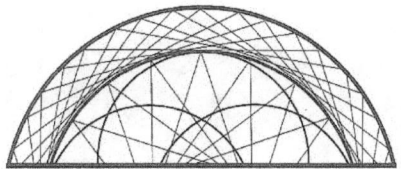

Chapter Four
Hidden Connections

With an equilateral triangle as a starting point, you'd think that extending the geometry into a larger pattern would be a simple matter. Not so. I tried extensions of all three sides. I tried encompassing the triangle with a circle. I tried overlaying a second triangle to create a Star of David. All drew a blank.

The line from St. Mary's, Monken Hadley to Camlet Moat went directly to St. John's Church in Clay Hill, Enfield, but beyond that there were no obvious mark points. Similarly, the line from Camlet Moat to St. Mary's, East Barnet could be lengthened as far as St. James' Church beside the golf course in Friern Barnet, but seemingly no further. The other patterns turned up nothing particularly relevant at all.

After several weeks of frustrated effort with no sign of any further developments, I paid St. Mary's in East Barnet a visit. As the Marian/White Goddess figure seemed to be the motivating force behind all this, I'd decided I should go along to the church and try some direct psychic contact with her.

The spiritual power station dream had suggested the site was a focus for a fountain-like outpouring of spiritual force traditionally described as the waters of life and as it was a key point in the Barnet Triangle I wanted to a clue how it connected with other places. After all, my own efforts in this direction were getting me nowhere.

At St. Mary's, I walked quietly around the churchyard mentally visualising an image of the white lady and focusing on the questions I wanted answered. I didn't expect an instant reply. By now I had realised that any communications were more likely to come indirectly through dreams, hunches, intuition or some synchronicity.

As it happened, the answer was not long coming. A few nights later, during a meditation, an image of the lady appeared, first at the centre of a five-pointed star, then at the centre of an eight-pointed star.

The images reminded me of the Tarot Major Arcana card called The Star, which actually shows a female goddess figure pouring the waters of Aquarius upon the Earth. I felt that it was a confirmation of the meaning of

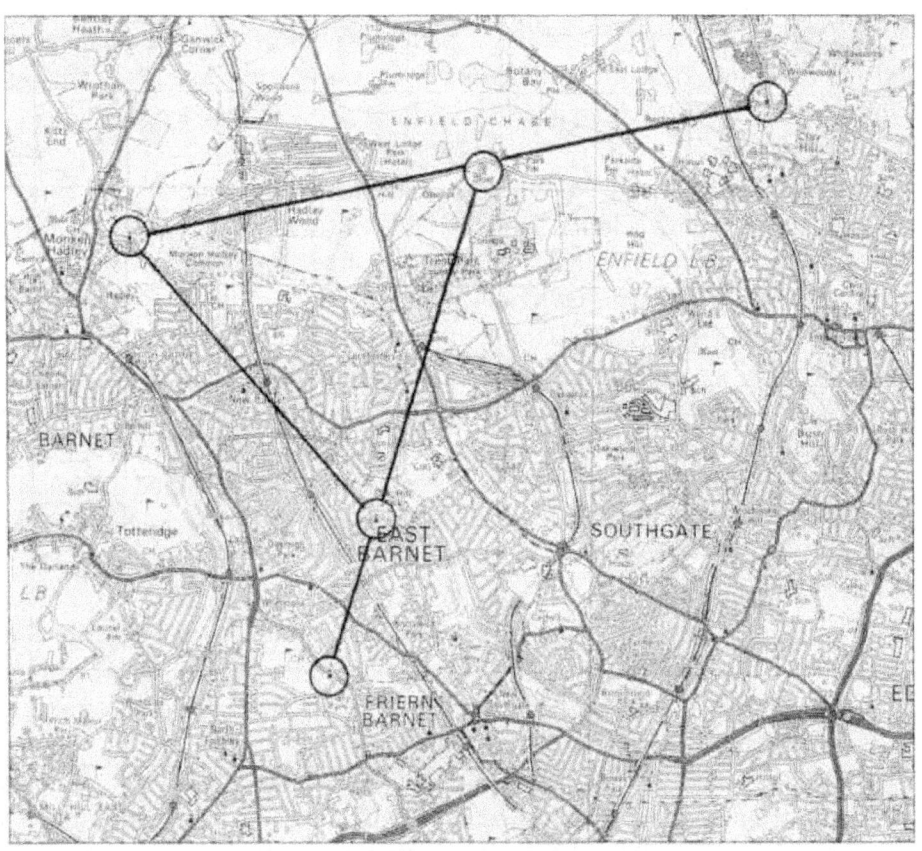

Illustration 15: The Barnet Triangle with local extensions to St. John's in Clay Hill and St. James' in Friern Barnet . O.S. map © Crown Copyright. All rights reserved. License No. 100029558.

the first dream and that the two star patterns were the next two geometric figures I should be looking for. Unfortunately, this knowledge didn't make them any easier to find. Neither pattern contains an equilateral triangle and my attempts to reconcile these disparate elements by following logical lines of thought came to nought. Then, once again, something other than logic lent a hand.

Zho Fritzler, a friend from the advertising agency I worked in, called to ask if I'd come with her to look over a building in Great Portland Street, a disused hospital. Someone she knew was interested in taking it over to use as an office, but they thought it had a strange atmosphere and, knowing my mystical interests included leys and their unusual energies, Zho wanted to see what I made of it. I met her at the building and while Zho spoke to the agent, I had a good look around on my own.

A deserted hospital is bound to have a bit of a strange atmosphere and this one certainly did. On the third floor, towards the rear of the building in particular, there were distinctly unusual sensations. In places, it felt like a warm breeze, or a very mild electric current. It wasn't in any way disturbing. On the contrary, it felt intuitively like a healing influence. When I looked at it clairvoyantly, I got a clear image of the Marian/Mother Goddess/Empress figure with the floral dress who I associate with St. Mary's in East Barnet. Was she trying to tell me there was some connection between the two locations?

When I rejoined Zho later, we stood outside the main entrance in Great Portland Street discussing our impressions of the place. As I told her about the Goddess image with the floral dress and the feeling of a connection to St. Mary's East Barnet, Zho looked down. There between us on the pavement was a bunch of Freesias. We were both absolutely certain that they hadn't been there a few moments before. If they had, surely we'd have noticed. If someone passing by had dropped them, we'd have definitely noticed. They would have had to walk right between us. Where the flowers came from was a complete mystery. If it was a sign from the spirit realms, it seemed a very good one. Even if I hadn't come up with anything significant so far, I felt it was meant to re-assure me that I was on the right track.

When I returned home, I checked my maps to see if there was any evidence to support the clairvoyant clue I'd been given, hinting at a link to St. Mary's in East Barnet. On the map, I lined up a ruler between St. Mary's and the old hospital in Great Portland Street. It passed through several distinct mark points, the first being ; St. Joseph's Church on top of Highgate Hill.

In recent years the church has been used as a chapel to Highgate School, but originally it had served as the local parish church and had been dedicated to St. Michael until the building of the current St. Michael's nearby in South Grove.

Since the church occupies the highest point in London, it would almost certainly have been a place of great significance to our ancestors who, by all accounts, equated high places with the abode of the gods. So, like St. Mary's, East Barnet, it could be a sacred place of impressive antiquity with connections to the druids. In THE BOOK OF DRUIDRY, Ross Nicholls, a former Chosen Chief of the Order of Bards, Ovates and Druids, has this to say about the area;

"The Grove at Highgate, Coleridge's home, was very probably a name marking the local oak or yew plantation of a Druidic College in this locality around the pool or spring of the goddess."

A closer look at the alignment in this vicinity showed it passes directly through South Grove and Pond Square. The pond may once been the pool or spring of the goddess, to which Ross Nicholls referred. Sadly, it has been drained and paved over.

Following the line south, it hit the notorious catacombs of Highgate Cemetery and a curious assortment of other places; the Victoria Memorial Fountain outside Buckingham Palace; Westminster Cathedral in Victoria Street; St. George's Church at Nine Elms; St. Leonard's in Streatham (another site with a history of religious use dating back a thousand years and on the brow of a hill); then, last but not least, Pollard's Hill in Norbury.

I have to admit that anyone deliberately looking for ley alignments on an O.S. map would probably never noticed this one at all. On face value, it just doesn't appear to have enough valid mark points over its 16 miles to qualify as an acceptable ley

However, I wasn't just looking for alignments on a map. I been led to this discovery by some very unusual events and without them, I wouldn't have been looking for anything. Because of that, I spent a great deal of time, checking out the alignment on the ground. For one thing, I was curious to know how the connections might develop from here and, more obviously perhaps, as this alignment passed through London rather than unspoilt countryside, I suspected that quite a lot of its original mark points may have been built over, so I wanted to see for myself what was there. Boy, was I in for some surprises.

Illustration 16: The north - south alignment. O.S. map © Crown Copyright. All rights reserved. License No. 100029558. The list below includes local landmarks as reference points as well as more significant mark points:

St. Mary's, East Barnet;
The grounds of Halliwick Hospital, Friern Barnet; Crossroads at Colney Hatch Lane and A406 (ancient crossroads were accepted ley mark points according to Watkins); Recreation Ground at Copett's Rd, East Finchley;
Former site of St. Thomas's Church in Creighton Rd, Muswell Hill (now demolished and replaced by modern housing);
The playing fields of Fortismere School, Muswell Hill (here the line crests the hill and exits through the school gates into Twyford Avenue);
Highgate Woods (the most obvious mark point here is a former sports pavilion, currently in use as a cafe);
St. Joseph's Church and school at the top of Highgate
Hill; Highgate's Pond Square;
The United Reform Church, South Grove, Highgate;
Highgate Cemetery, in particular, directly through the notorious Catacombs;
Crossroads at the bottom of Highgate West Hill and Chetwynd Rd, near Royal Oak, flanked by two churches, Highgate Baptist Chapel and a Greek Orthodox Church in Gordon House Rd;
The Church of The Most Holy Trinity with St. Barnabas in Clarence Way, NWI which looks a little neglected but has a powerful atmosphere; Community gardens (between Clarence Way and Leybourne Rd, Camden) which provide some welcome greenery for the area; The line crosses Chalk Farm Road near Camden Lock Markets; Our Lady of Hal, Arlington Rd, Camden; St. Bede's Chapel, Albany Street, now in use as a gym;
Direct alignment with the Marylebone Rd end of Osnaburgh Street NWI where the alignment passes the western end of St. Mary Magdalene, Munster Square and the eastern side of Holy Trinity Church-for many years used by the Society for the Promotion of Christian Knowledge; The Central Synagogue, Hallam Street;
Site of a former St. Paul's church at the junction of Gildea Street and Hallam Street-Oxford Circus (another crossroads); Green Park (circle of trees);
Victoria Memorial Fountain in front of Buckingham Palace;
Westminster Cathedral;
The Church of St. George and St. Andrew, Patmore Steet, Nine
Elms; St. Peter's Parish Church, Clapham Manor Rd; St. Leonard's
Parish Church, Streatham; St Bartholomew's Catholic Church,
South Streatham; Pollard's Hill, Norbury.

City of

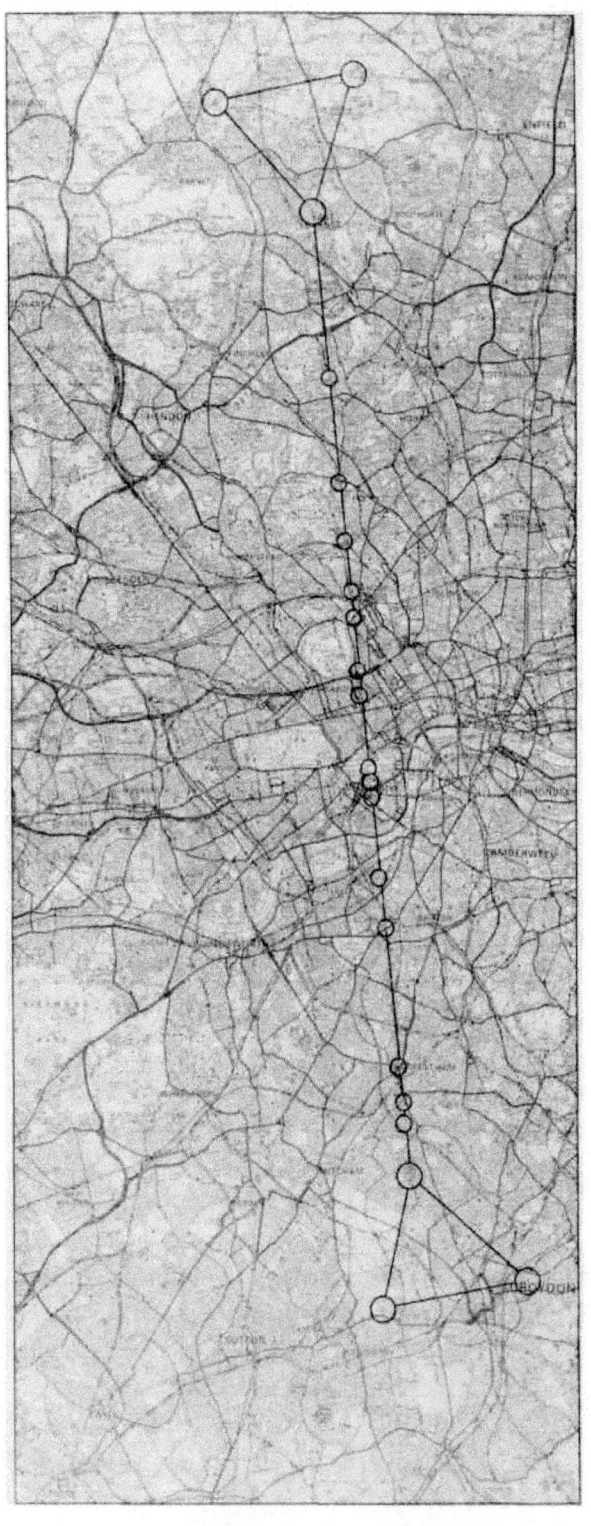

Illustration 17:
St. Joseph's, Highgate, on the summit of Highgate Hill, the highest point in London.

Illustration 18:
The splendid Queen Victoria Memorial Fountain outside Buckingham Palace.

GREEN PARK'S GROVE

Surprise number one was Green Park. Where the line crosses the Eastern side of the park near Spencer House there's a perfect circle of trees. If you walk directly into the park from the tube station or along Queen's Walk, it's the first large clump of trees you'll see. The line doesn't go through the middle of the circle. It forms a tangent to it as if either the circle or the line might pick up a spin of energy from each other. Some of its strange vibes are less esoteric in origin. They are the rumblings of Victoria Line tube trains which pass beneath this part of the park.

I was astonished enough to find the trees formed a perfect circle, but when I counted them, I was even more astonished. There are 13. One tree for each of the 13 full moons or lunar months in the year and 13 sacred trees of the druid calendar. This is a perfect druid grove.

To add to the enigma, the trees are reasonably evenly spaced, but for one pair on the Queen's Walk side. These two have a larger gap between them. By coincidence (or possibly not) this gateway faces to the North East and the midsummer sunrise, just like the axis of Stonehenge's inner circles. Obviously, a midsummer sunrise would never have been visible from this circle. The sight line has been built upon and blocked for quite some time, probably even before these trees were planted. Nevertheless, for reasons destined to remain inexplicable, that's the direction identified.

Who was responsible for planting this circle of trees remains a mystery. All The Royal Parks Authority could tell me is that they were planted during the last century and that, for some time, a bandstand had stood at their centre. Supposedly, it was removed in the 1970's after it became a popular spot for couples to enjoy naughty nocturnal adventures.

My feeling is that this grove was deliberately planted for use by some local mystical group or secret society whose leading members lived nearby. When visiting the circle, on Halloween 1992, I clairvoyantly saw a procession of white-robed figures entering the circle from the Queens' Walk side to perform some kind of ceremony.

It is certainly a place of power and an extremely pleasant place to enjoy an outdoor meditation or simple ceremonies. The atmosphere when you enter the circle is often quite magical and so calm that you easily forget that you are in central London a few yards from one of its busiest main roads.

Illustration 19: Green Park's mysterious druid grove. A perfect circle of 13 trees, one for each of the 13 lunar months (There are actually 13 full moons in our 12 month year).

WESTMINSTER CATHEDRAL

From the grove in Green Park, the alignment passes directly through the Victoria Memorial Fountain with its impressive golden angel, in front of Buckingham Palace.

The next major mark point is Westminster Cathedral in Victoria Street. The alignment passes diagonally through the Cathedral, exiting the building in the area of the Lady Chapel on the south west side.

Initially, I was a little disappointed that the alignment had struck Westminster Cathedral rather than the more famous and rather more ancient landmark of Westminster Abbey. The Cathedral's foundation stone was laid by Cardinal Vaughan in 1895, over a thousand years after Westminster Abbey had opened for business and its immediate predecessor on the site was the New Bridewell penitentiary, so it didn't seem to have much heritage as a sacred site. Nevertheless, a little historical delving soon revealed two facts that may indicate otherwise. Not only does it stand on land which once belonged to the Benedictines of Westminster Abbey, the open forecourt between the Cathedral and Victoria Street was once a field where an annual fair dedicated to Mary Magdalene was held every July 22nd, her feast day. Such a fair, especially with its specific dedication, could indicate that the site may have been the focus of ritual or ceremonial use for centuries.

LONDON'S HILL TOP HENGE

Another great surprise was Pollard's Hill. It's marked on the Ordnance Survey maps simply as a high point with panoramic views and an O.S. triangulation point, so I had no inkling what was actually there, apart from a hill, and I didn't immediately feel inclined to visit the place. Then, one Sunday morning, I had another of those psychic wake up calls. I found myself wide awake uncharacteristically early, with an overwhelming compulsion to go there. So I did.

You can reach Pollard's Hill from a turning off the main A23 London to Croydon road. The area is dominated by a relatively pleasant housing development dating from around 1930. Driving through it hardly prepares you for anything out of the ordinary, yet all this suburban normality hides something quite unexpected, masquerading as a small park. I left my car by the iron gates at the bottom of Pollard's Hill and walked casually to the top. On the way up, it looked like a pretty average hill. The views in most directions became spectacular, but not as spectacular as what I found beneath my feet at the summit.

Illustration 20: Westminster Cathedral may not be as old as Westminster Abbey, but the land was owned by Westminster's Benedictines and was the site of annual festival or fair dedicated to St. Mary Magdalene held every July 22nd, her feast day. The tradition suggests it was an ancient ritual site.

The crown of the hill is an earthwork circle, like the remains of a henge. Standing in the centre was like being in a huge grassy crater which had all the atmosphere of Stonehenge, but no stones.

The highest point was the northern side and could have been the remains of a mound or barrow. To the south was a smaller mound near what could have been an entrance through the earthwork banks. Clairvoyantly, I could see an image of a single trilithon or dolmen arch on the northern crest if the hill. If it had ever stood here in reality, like so many of this country's megalithic monuments, the stones had long been looted to provide building materials.

Despite the fact that this is obviously an important ancient site, the local council had seen fit to erect a hideous shelter on the north east banks of the earthwork. Its most frequent users, by the look of it, appear to be the local graffiti artists whose efforts have made it even less attractive.

I picked up quite a lot of psychic impressions at this place. First, as I walked up the hill, an image of the White Goddess was apparent as Bride, wearing a white dress and holding a bunch of flowers at waist level. I was also aware of the image of a Druid site guardian, full bearded, holding a staff and with a large crescent-shaped breast plate around his neck, made of gold.

Several thoughts came to mind in connection with this figure. He seemed to identify himself with 'The King' who he said was buried beneath the northern mound and with whom he had a close association during his lifetime. He himself had been the Chief Druid at the site which, he said had been an important spiritual centre whose power derived from its nature and position. Interestingly, he didn't see himself as dead, but as existing in a different world to ours and available as a source of help and information to those genuinely seeking to be of service to the spirit of his land.

Two geometric patterns came to mind whilst tuning in at Pollard's Hill. One was a fourfold arrangement of circles. The other, a triangle mirroring the Barnet Triangle at the northern end of the alignment.

Both will be dealt with in greater detail later.

Illustration 21: The grassy crater and circular mounds atop Pollard's Hill in Norbury. It looks to me like a hill top henge and ancient ritual site of great significance (Photograph courtesy of Jonathan Trapman).

LONG DISTANCE INFORMATION.

Surprise number three was that the line from St. Mary's, East Barnet across London to Pollard's Hill could be tracked all the way to the south coast by way of a startling succession of hilltops, invariably occupied by sacred sites of great antiquity It took me a considerable time to visit them all, but it did confirm that I was not dealing here with some chance occurrence.

South of Pollard's Hill, the alignment goes directly to another hilltop, the summit of Russell Hill, now occupied by a school, but formerly another important druidic location, if my psychic insights were to be believed. The next mark point is another hilltop in Old Coulsden, crowned by St. John's Parish Church. According to the church's own notice board, there has been a place of worship here for over a thousand years. Next, came the old Castle Hill Mound at Bletchingly Access to this looked nigh impossible and its appearance was like a smaller version of Sillbury Hill covered in impenetrable undergrowth. St. Leonard's at Turner's Hill was the next mark point and another lofty hilltop, followed by St. Peter's at Ardingley

An odd man out is St. Michael's at Plumpton. Unlike the others, it's on low ground. It is, however, an exceedingly ancient site in common with the rest.

The line reaches the English Channel via Rottingdean and, looking at the map, it seems to miss the town's obvious mark points rather disappointingly. To the East of the alignment is the old Parish Church, St. Margaret's, where a place of worship has stood since Saxon times. On the west is a beacon hill. The alignment goes neatly between the two. From this, you might assume that Rottingdean holds nothing else of any significance. Wrong. The most important thing the alignment indicates here is the importance of fieldwork. What the map does not show is that Rottingdean retains the remnants of an ancient megalithic monument and several stones or various sizes can be found scattered around the town. Some are directly on the alignment and one, set into the road near North End House, close to the village green, had a distinct tingle to the touch on the day I visited. For good measure, there is also an old well on the village green, directly on the alignment, so it's very well defined on the ground, if not on the O.S. map.

To the north of London, the alignment goes directly to the old church of St. Nicholas at The Bury, near Stevenage. Beyond that, its

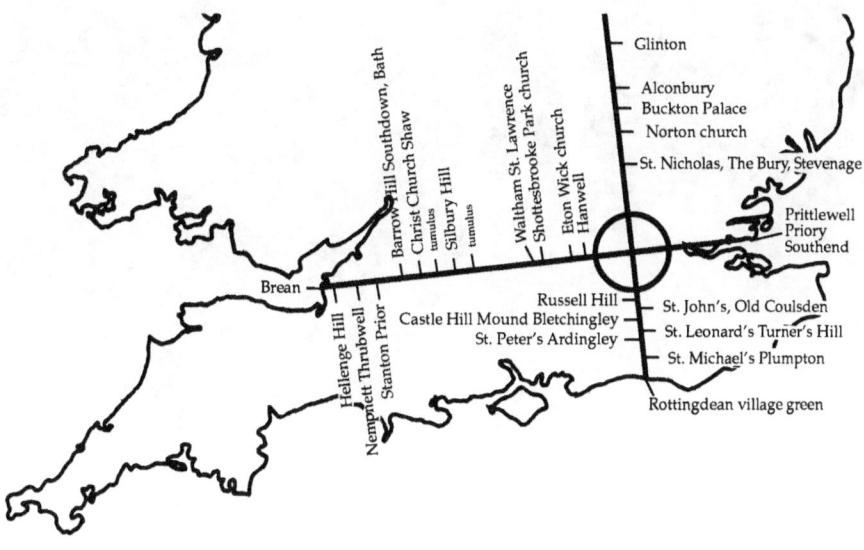

Illustration 22:
The north - south line is well defined, through a startling succession of ancient sites.

mark points are less distinct. If the alignment does extend further north, it would reach the North Sea in the area of Ness Point, Robin Hood's Bay (not far from Whitby Abbey).

The entire distance from St. Mary's in East Barnet to Rottingdean is over seventy miles. From St. Nicholas at Stevenage, it's nearer a hundred miles. For an alignment of that distance to be so clearly and accurately delineated with so many important sacred sites, the majority of which are hilltops, is obviously not just coincidence. I took it as evidence that I was definitely working along the right lines.

Illustration 23:
The north - south line reaches the south coast at Rottingdean, where it passes through this well on the village green. Stones from a megalithic monument can be seen nearby.

LINES OF SPIRIT

At various places along this alignment, I could clairvoyantly sense the same intriguing White Goddess or Marian presence; At St. Mary's in East Barnet, of course; in Highgate Wood; at St. Leonard's in Streatham; at Pollard's Hill; near the well on Rottingdean's village green. A similarity between these "Genius Loci" suggested that the idea of spirit of place should be interpretted in a broader context. Over a wider area such as this, it may understood as an aspect of the spirit of the land, or of the Earth.

What struck me as even more astonishing was that the presence of the Divine Mother was the focus of daily worship and reverence at two relatively modern sites. The first was a small church in Arlington Rd, Camden, dedicated to Our Lady of Hal.

Our Lady of Hal is inconspicuous. You could easily pass by without noticing it. Yet inside, it has the atmosphere of a spiritual power

Illustration 24: The Madonna and child statue on the alignment through Westminster Cathedral.

station and for me, the focus of power was not at the altar, but a shrine to the Virgin Mary, located near the entrance. As with many Catholic churches, the shrine has a life-size statue and the shock of recognition I felt when I saw it was almost overwhelming. The resemblance to the apparitions I had been seeing was uncanny, almost like meeting her in the flesh.

When I visited Westminster Cathedral, a similar experience was in store for me. The alignment cuts diagonally through the building and passes directly through a statue of the Madonna and child in the Lady Chapel. Here again, the statue seemed to possess a special power and presence.

Clairvoyantly, I could see a Mother and child but on one occasion, the name Isis Maria was picked up. Interestingly, some statues of Isis and her child Horus bear a remarkable similarity to those of the Madonna and child. To find later that the Cathedral stands on a ritual site with a dedication to Mary Magdalene was, for me at least, confirmation that the spirit of place was (and always had been) that of a beneficent female presence. It suggests a common, archetypal influence behind the pre- and post-Christian Holy Mothers and when

I mentally asked about this, I was given a clear message that; **"Both were the Queen of Heaven of their time and in the minds of the people."**

In his book STATIONS OF THE SUN. the historian Professor Ronald Hutton points out a similar relationship between St. Bridget, The Mother Saint of Ireland and her pre-Christian counterpart.

" It seems reasonably certain that behind this alleged holy woman of whom no contemporary or near contemporary records survive, stands a pagan goddess of the same name. What is by no means clear is whether there was one goddess, or a triple one, or several, and whether there was in addition a real Christian woman of the same name with whom the deity became conflated."

Suggesting there could be some similarity between the Virgin Mary and a pagan goddess, albeit in terms of Jungian archetypes, could stir up all kinds of trouble, yet these experiences gave me the distinct impression that there was some kind of power or energy in both the alignment and its sites which could manifest and be seen as both the Christian Holy Mother and her pagan predecessors, as if the same source had evolved both from a common psychological archetype. I had also felt some unusual sensations and energy effects at various points along this alignment. As similar apparitions could be sensed at many of these locations, I was certain that the white ladies were manifestations of the energy present at certain places, spirits of place, localised aspects of the Earth Spirit and a component of the universal life force.

I also began to suspect that alignments of these apparitions constituted spirit pathways, somehow linked to the intrinsic soul energies of the land or planet as a whole. Within this broad concept, the alignment could be considered as a pathway of the Earth Spirit or an expression of the power of the Divine Mother, in all her forms. Equally, the power may be sensed simply as energy, rather than spirit, by those who lack a clairvoyant perspective on these matters. Dowsers frequently report detecting mysterious energies, both at ancient sites and flowing between them, although invariably, even those who refer to them as Earth energies, acknowledge that they have a spiritual basis.

At this point, I have to say that I should take no credit for originating the concept of the Earth Spirit as Divine Mother. The seed of the idea had been planted some years before when I read a book by John Michell entitled THE EARTH SPIRIT, ITS WAYS, ITS SHRINES, ITS MYSTERIES. In it, John points out that these beliefs were once commonplace and it is only in the last few centuries that they ceased to become the accepted norm.

"The orthodox view that has survived into the middle ages from prehistoric times is expressed by the alchemist Basilius Valentinus; "The Earth is not a dead body, but is inhabited by a spirit that is its life and soul. All creatures, things, minerals included draw their strength from the Earth Spirit. This spirit is life. It is nourished by the stars and gives its nourishment to all the living things it shelters in its womb."

"The Earth was sacred, not because pious people chose so to regard it, but because it was in fact ruled by spirit, by the creative powers of the universe, manifest in all the phenomena of nature, shaping the features of the landscape, regulating the seasons, the cycles of fertility, the lives of animals and men. In this secure immortal world, the most assured reality was communication with the local gods, personifying aspects of the universal spirit of the Earth."

That latter point is the key to the comprehension of this phenomena; to our ancestors, the world was alive with spirit; the spirits of nature; the spirits of the ancestors; the spirits of the land, trees and stones; the spirits of the regional gods, or local mythological heroes.

In this enchanted realm, the genus loci of individual shrines may be understood as a localised manifestation of the spirit of the land, which in turn, in a holistic universe, is an intrinsic part of the universal spirit, the creator. It is a simple logical progression to regard alignments of shrines dedicated to similar deities as some kind of spirit paths. The idea of deities travelling these spirit paths is equally understandable if the local god, goddess, spirit of the land, or whatever are personifications of natural forces that are directed along such alignments. They are lines of both energy and spirit. In fact, this is how many native societies regard them.

It is easy for those accustomed to modern city life to take a cynical, disbelieving stance of this world-view since they may never personally experience the reality of it. By contrast, it was rapidly becoming part of my day-to-day reality. Only from this perspective did these experiences and discoveries made any sense. I had no choice but to accept it.

THE CROYDON TRIANGLES

On the first visit to Pollard's Hill, I sensed that there were connections to other local sites which may form a triangular relationship similar the Barnet Triangle at the northern end of the line.

It's hard to explain how ideas like these form, or how any ideas form for that matter. Where does inspiration come from? Robert Graves attributed it to the White Goddess in her role as the Muse. I don't doubt it. At Pollard's Hill, as elsewhere, the clairvoyant image of the lady in white did seem to be the source of these notions.

As I stood atop Pollard's Hill taking in the breathtaking views, the image of the white lady appeared, then visibly receded into the distance, towards the south west. Clairvoyantly, I saw a flash of her standing in a churchyard. There seemed to be trees and grassland nearby and the impression of a small bridge over a stream.

I had an Ordnance Survey map with me and could see immediately that a church to the south west at Beddington might be the location. When I visited it later that day, I found it was another St. Mary's with an impressive history, having been a place of worship since Saxon times. Like St. Mary's church in East Barnet, it stood on the fringe of a large and pleasant park which had probably helped preserve its special atmosphere from the epidemic of uncontrolled building which has afflicted our countryside over the last century.

To complete the triangle though, there had to be a similar site to the south east and while I could sense this from Pollard's Hill, it wasn't as easy to identify as St. Mary's at Beddington. For one thing, the first site that attracted my attention was much too close to create an equilateral triangle like Barnet's, but puzzlingly, it produced a positive reaction to map dowsing with a pendulum.

A visit revealed it to be the old parish church in the centre of Croydon, dedicated to St. John The Baptist. Again, it's a church with an extensive history. The first records date back to 960 AD. Until as recently as the last century, the church had stood upon an island, so if it had evolved from a pre-Christian shrine, it may have been some kind of sacred isle or enclosure like Camlet Moat. It had a direct correspondence with the moat as home to the local manor house, too. Next to the church on the isle had been the Palace of Croydon.

A Sunday morning service was in progress when I reached St. John's. This did allow me access to the church, but I don't recall picking up any

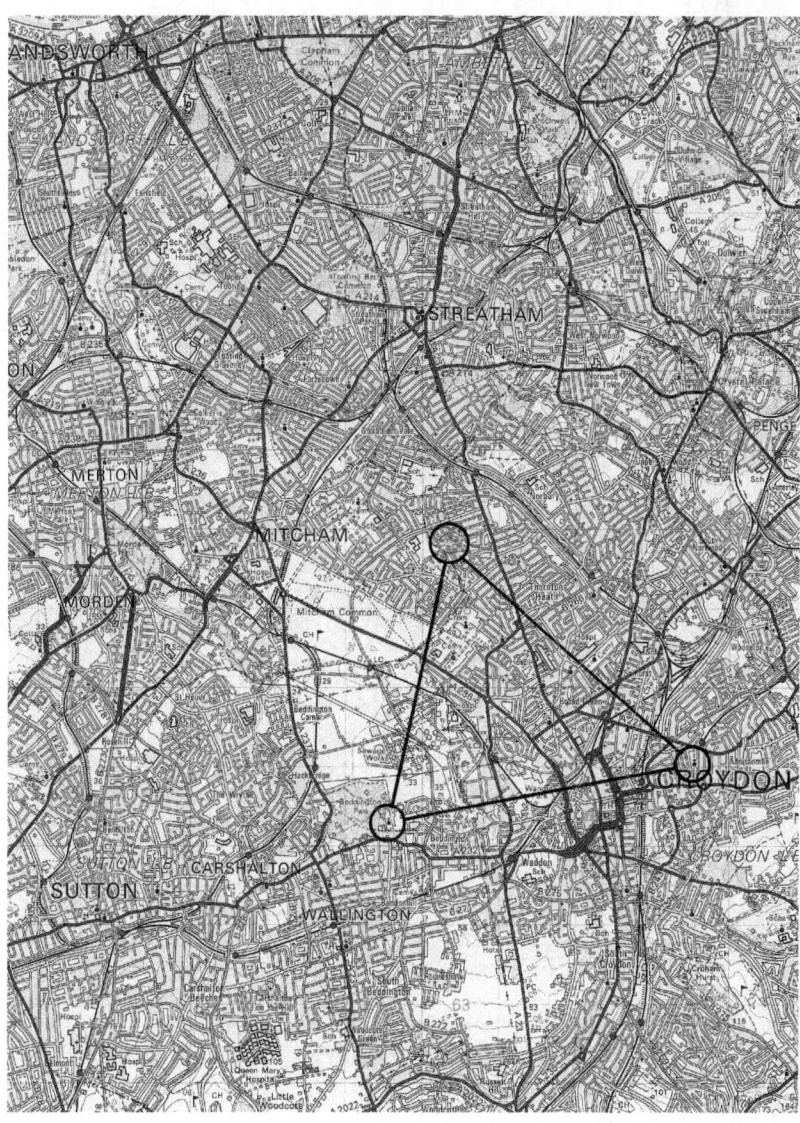

Illustration 25: The Croydon Triangle, St. Mary's Beddington, St. Mary's Addiscombe and Pollard's Hill, Norbury. O.S. map © Crown Copyright. All rights reserved. License No. 100029558.

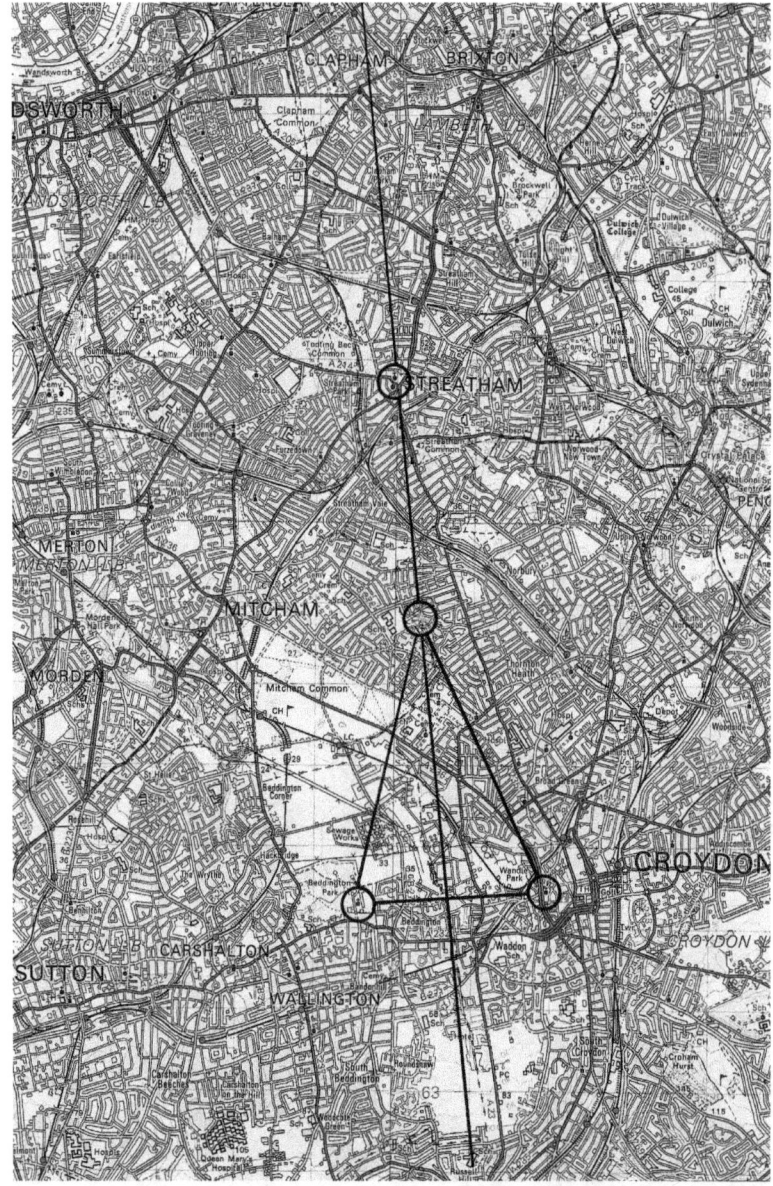

Illustration 26: Croydon's isosceles triangle pointing at Pollard's Hill and on up the line, formed by Pollard's Hill, St. Mary's, Beddington and Croydon Parish Church of St. John the Baptist. O.S. map © Crown Copyright. All rights reserved. License No. 100029558.

strong clairvoyant images here, possibly because of the distractions caused by the large numbers of people in and around the church as the service ended.

The second site I'd been drawn to, and which map dowsing had identified, turned out to be St. Mary's in Addiscombe. It was by no means ancient, a classic example of Victorian Gothic. Yet surprisingly, the presence of the White Goddess could be sensed quite clearly.

When I returned home later that day, I continued work on a large scale O.S. map. The triangle formed by Pollard's Hill, St. Mary's Beddington and St. John's Croydon was much smaller than Barnet's. At the base of the N-S line, it had the curious effect of looking like an arrow pointing up the line, or positioned to identify Pollard's Hill.

Superficially, it looked as if it could be an isosceles triangle, with lines from Pollard's Hill to each to the churches of equal length. A quick check with a ruler soon revealed that they were almost equal, but not quite. I turned my attention to the second triangle. This wasn't equilateral either, but it had two St. Mary's, both identified by the White Goddess herself, St. Mary's Beddington and St. Mary's Addiscombe, so I marked it on my map to see what might develop from it as the discovery progressed.

CROYDON TRIANGLE DIMENSIONS

Pollard's Hill to St. Mary's, Addiscombe;
About 2.625 miles or 21 furlongs.

Pollard's Hill to St. Mary's; Beddington;
About 2.25 miles = 18 furlongs or 3960 yards or 720 poles.

St. Mary's, Beddington to St. Mary's, Addiscombe; About 2.516 miles.

Perimeter of triangle; 73.9 miles approximately.

A CROP OF CIRCLES

The alignment from St. Mary's East Barnet to Pollard's Hill cuts clear across London, right through the centre. As I hadn't found any trace of a five or eight pointed star so far, I decided to use this alignment as the diameter of a circle, to see if it threw up any new leads.

The mid-point of the line turned out to be just to the north of Oxford Circus near the junction of Hallam Street and Gildea Street, roughly midway between the Central Synagogue and All Soul's Langham Place. Taking a centre here and a radius to St. Mary's East Barnet or Pollard's Hill, creates a huge circle which encloses almost all of Greater London. Even as I drew the circle, I could see that it passed through a large number of interesting spots, including Horsenden Hill at Greenford and Caesar's Camp on Wimbledon Common.

Encouraged by this result, I drew a second, even larger circle, using a radius to Camlet Moat. I was amazed to find that at the southern end of its sweep, the circumference passed through St. Mary's Beddington, justifying its inclusion as part of the Croydon Triangle.

Why should these places, twenty miles apart and separated by some of the most dense urban development in the country, be linked by such an unusual connection? I had no idea, but I felt it did confirm that my intuition had been correct to regard the Croydon Triangle sites as a relevant part of the emerging pattern.

INNER CIRCLE DIMENSIONS

Diameter;
 518.4mm on OS 1;50,000 map = 25.92km = 16.2 miles = 129.6 furlongs
= 28,512 yds = 85,536ft = 1,026,432 ins = 5,184 poles or lugs
Circumference;
50.9004 miles approximately.

OUTER CIRCLE DIMENSIONS

Diameter;
660mm on OS 1;50,000 map = 33km = 20.625 miles = 165 furlongs =
36,300yds = 108,900ft = 1,306,800ms = 6,600 poles or lugs.
Circumference;
129.6 miles approximately 1036.8 furlongs =228,096 yds =
684,288 ft = 41,472 poles or lugs.

Illustration 27: The two concentric circles linking the Barnet and Croydon triangles. O.S. map © Crown Copyright. All rights reserved. License No. 100029558.

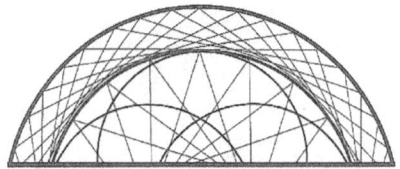

Chapter Five
The First Earthstar

The maps I had been using to plot these two concentric circles were Ordnance Survey 1;50,000. Unfortunately East and West London are on separate sheets so to cover the whole of Greater London, I had to trim the maps and mount them on a huge piece of art-board. The size of the map meant I then had to invest in a beam compass capable of drawing circles up to 28" in diameter. It all took a considerable amount of time, effort and expenditure, but it was well worth it.

Even as I drew in the circles for the first time, I could see that some interesting spots lay on their perimeters. One of the most prominent sites on the inner circle (and on the landscape at large) was Horsenden Hill in Greenford. I laid a protractor on the map and, from the centre of the circles, measured the angle between Horsenden and St. Mary's in East Barnet. It was exactly 72 degrees. 72 degrees is precisely one fifth of a 360 degree circle and a pentagonal angle, so I immediately realised that this could be a clue to the five-pointed star I'd been looking for. Sure enough, another 72 degrees anti-clockwise from Horsenden Hill was a point near Caesar's Camp on the edge of Wimbledon Common. With these three sites established, it was easy enough to plot the locations of the fourth and fifth points of the star.

Exactly 72 degrees clockwise from St. Mary's East Barnet was a church opposite Wanstead Flats, St. Gabriel's. 72 degrees further on was the final point near Bellingham Green, which has a geometric pattern of its own. Bellingham Green is roughly hexagonal, contains some mysterious-looking mounds and is flanked by two churches, the older dedicated to St. Dunstan.

At first, I had my doubts about whether this was a valid figure because, although the 72 degree angulation is correct, Bellingham Green and Caesar's Camp are close to the exact geometric points rather than directly on them. The precise geometric spot at Bellingham is on the edge of playing fields nearby. The exact geometric point on Wimbledon Common is near a spring in woodland a little to the south of Caesar's Camp. The more I explored this pattern, though, the more features I found to convince me that it is built upon a very real foundation. First, there are the star's radial axes. When drawn as five diameters of the circle, these

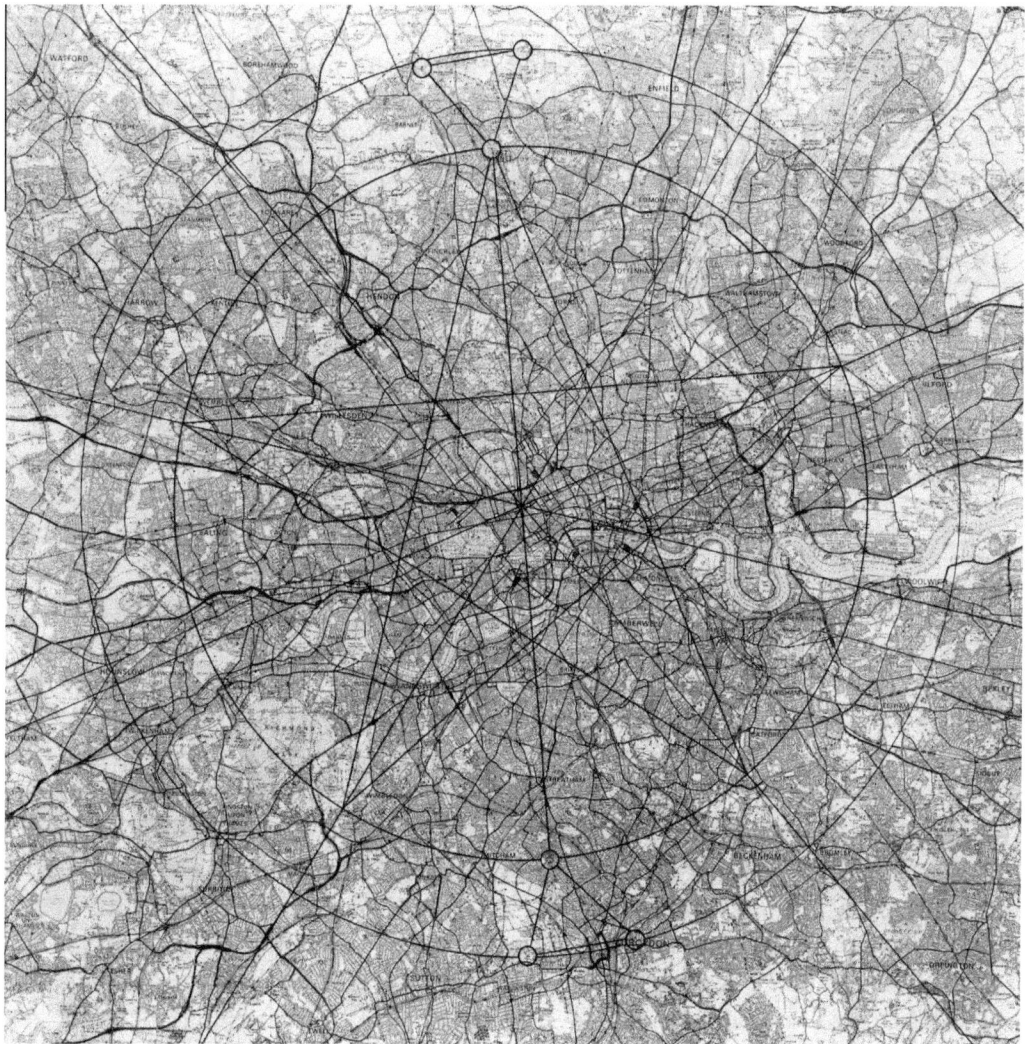

Illustration 28: The huge pentagonal star concealed within London's landscape. O.S. map © Crown Copyright. All rights reserved. License No 100029558.

define ten points on the circumference, not just five. I also extended them out beyond the circle of the city, to see what other sites, if any, they picked up. Even before I'd begun to look for likely mark points along these lines, something astonishing stood out on the map. Every single one of them runs parallel to one of the major roads converging on London.

On closer examination, these main arterial routes hold a significance of their own. They predominantly follow the path of Roman roads and so are of great antiquity. Some sources suggest that these roads actually pre-date the Romans who merely utilised existing British roads and trackways built by the British Kings Moelmutius and Beli Mawr, father of King Llud, so their origins may be very ancient indeed.

They are, without doubt, some of London's oldest travellers' ways and large stretches of them are obviously aligned to the same directions as the pentagonal star's five axes.

Let's start with the A40. Running alongside it is the axis from the centre of the star through Horsenden Hill. In the opposite direction, to the south-east, the same axis marks a course perfectly parallel to the A2 all the way from Greenwich through Bexley and Dartford into Kent.

Out in Essex, the A12 and the A112 through Ilford, follow the course of an old Roman Road. The star's north-east axis through Wanstead runs alongside it from as far out as Brentwood.

On the opposite side of London, this same line runs parallel to the arrow-straight A30 to Staines, another Roman Road. In fact, this alignment tracks the road all the way from its beginnings as Notting Hill Gate and Holland Park Avenue.

The old Roman Road of Watling Street, now the A5 Edgware Rd, runs alongside and almost parallel with another of the star's radial axes, which runs north-west to south-east down to the Bellingham Green point. The same line also tracks the course of the A41 Finchley Rd into London, with several parallel sections.

Finally, the main north-south axis of the star matches the course of the A23 Brighton Road for several miles south of Croydon, while to the north, it overlays the meandering A1 all the way up to Huntingdon and beyond. The obvious question is why ? Why should Roman roads follow precisely the same direction as these mysterious star lines which are not even visible on the landscape? Were our original roads laid out to the same vast groundplan?

Frankly, I don't know. But it does suggest that the roads and the star pattern may both be connected to the same mystery, concealed within the landscape.

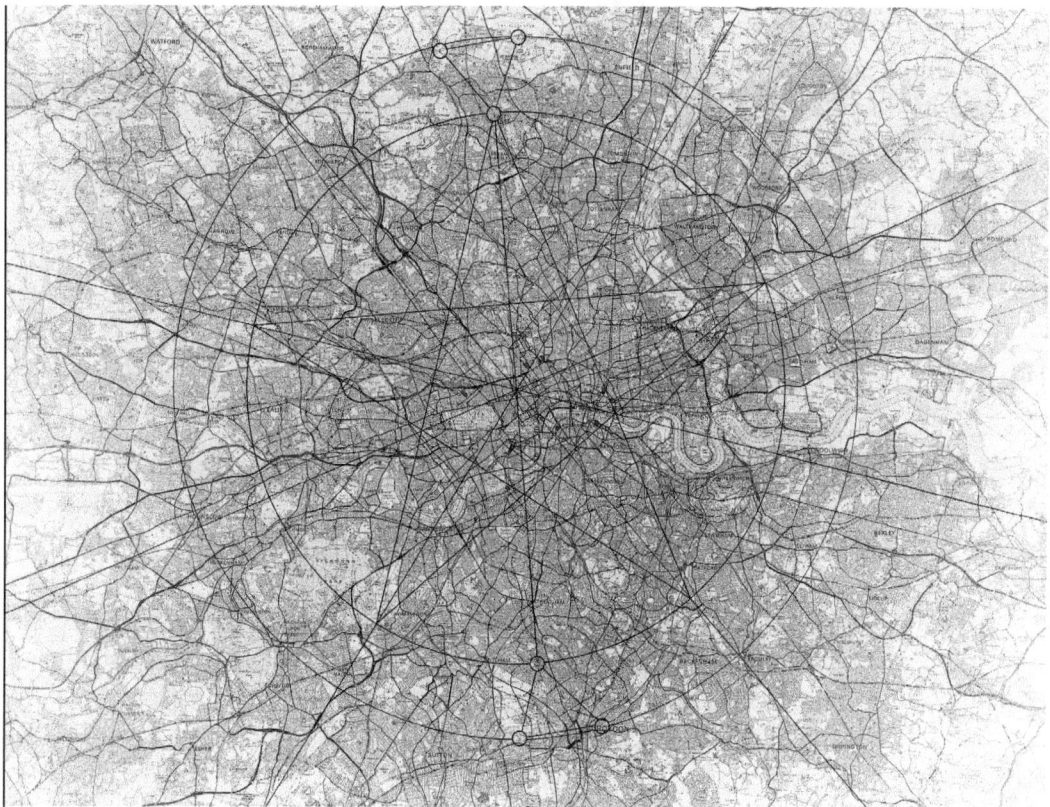

Illustration 29: The capital's ancient main roads run parallel to the star's axes as if both were laid out to a common plan. O.S. map © Crown copyright. All rights reserved. License No. 100029558.

PENTAGONAL MARK POINTS

With straightforward ley alignments, four or more mark points over the course of a couple of miles is enough to stack the odds against them being chance alignments. Obviously, with a pattern of this sort, things get a little more complicated. If the alignments are all well-defined by important sites as well as forming part of a complex piece of geometry, the chances of any of it being something which has occurred by accident would be pretty slim.

With this in mind, I looked very carefully at the constructional lines of this pentagonal design both on the maps and on the landscape. The results were very encouraging. As well as the strange correspondences with London's main roads, some of the alignments had an astonishing number of mark points, many of them places or sites of immense importance in London's history.

THE HORSENDEN HILL AXIS

The axial line through Horsenden Hill, for instance, strikes a startling succession of sites, including at least six hill tops, a Neolithic camp, a holy well, the site of central London's only remaining megalithic stone, at least seven churches, five of them of considerable antiquity, and last but not least, a square moated islet similar to Camlet Moat.

With an excess of twenty mark points, this alignment could not possibly be described as the product of coincidence.

As if to emphasise the point, many of its sites are of exceptional interest in their own right.

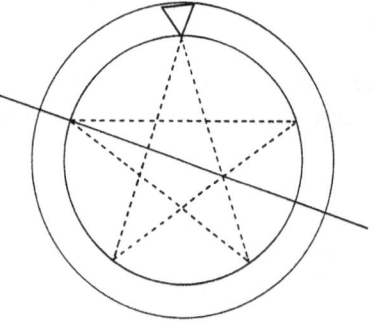

tration 30: Diagram of the Horsendon Hill axis.

MARK POINTS ON THE HORSENDEN HILL AXIS.

I: Bulstrode Camp, a Neolithic earthwork on a hilltop near Gerrard's Cross.
2: St James' Church in Gerrard's Cross.
3: St. Mary's at Denham, a church of great antiquity.
4: St. Giles at Ickenham where the line actually passes down the road at the side of the church and through the site of the old market cross at the nearby crossroads.
5: The summit of Dabbs Hill which is also on the outer circle.
6: Another hilltop near Lancaster Rd, Northolt, currently used as an Air Traffic Control radio station.
7: Horsenden Hill in Greenford. Like Pollard's Hill, it is a prominent vantage point from which you can see for miles in every direction.
8: One Tree Hill in Alperton, another prominent hill with an evocative name and mystical atmosphere.
9: Church in Harlesden.
10: Curious parallel to suburban roads in Kensal Green.
11: The centre point of the star, near the site of a former St.Paul's Chapel at the junction of Gildea Street and Hallam St, WI.
12: The Law Courts in Fleet Street, designed by G. E Street.
13: St. Dunstan's, Fleet Street.
14: St. Bride's, Fleet Street.
15; The line follows the course of Cannon Street passing south of St. Paul's and running close to the course of yet another old Roman Road, Watling Street.
16: The site of the London Stone in Cannon Street, central London's sole remaining megalithic stone.
17: Trinity Gardens overlooking the Tower.
18: Tower Hill.
19: Bostall Woods.
20: Ancient moat at Howbury Farm, Slade Green.
21: Alignment parallel to more than 12 miles of the A2 in Kent.

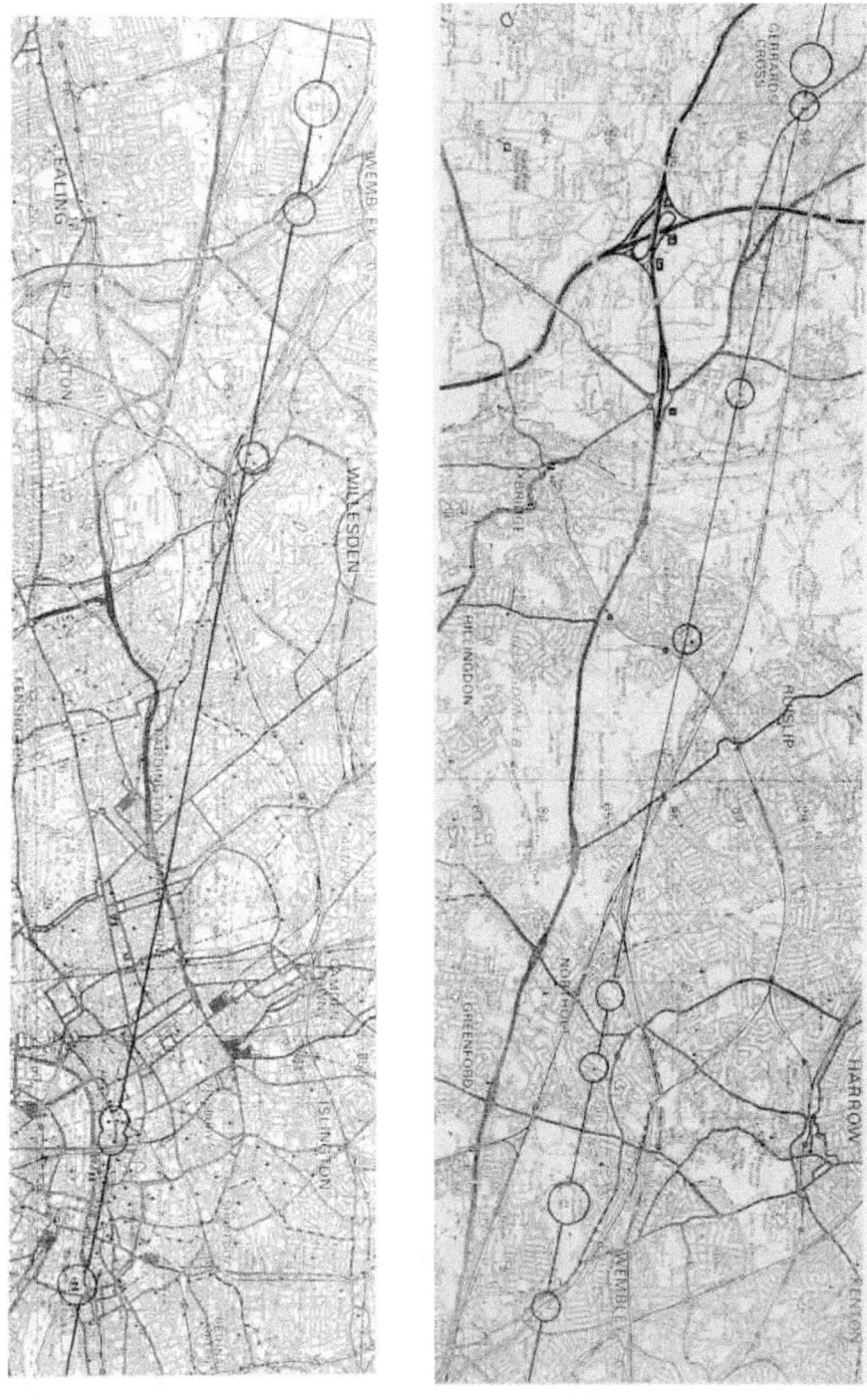

Illustration 31: The Horsenden Hill axis of London's five-pointed star; Mark points listed on page 74 (O.S. map © Crown copyright. All rights reserved. License No. 100029558).

THE LONDON STONE

In the whole of central London, there is only one surviving ancient megalith, the London Stone. This is the only remnant of the capital's megalithic monuments and although no longer in its original position, the Horsenden Hill axis passes directly through the site where it once stood and is wide enough to also take in its present location, near to Cannon Street Station. This is extremely significant. The London Stone may have only survived because it was the most important part of London's megalithic complex, the omphalos stone of the capital.

An omphalos is the primary foundation stone of any temple and the most significant for the simple reason that it is imbued with the spiritual power of the entire site and surrounding area. Nigel Pennick's description of it in THE AQUARIAN GUIDE TO LEGENDARY LONDON states;

"According to tradition, the London Stone was originally a temple altar stone laid by none other than Brutus the Trojan, mythical founder of Britain. "

Through some on-site mediumship, I learned that the temple's other stones were removed and broken up to build London's city walls. Only because of its importance was the omphalos stone spared. It is possible that no-one dared destroy it. Throughout the centuries it has been connected with the preservation, safety and well-being of the capital. An old saying goes;

" **as long as the stone of Brutus is safe, so long shall London flourish."**

Despite this, very little of the stone has actually survived. Today, its remains are preserved in a pitiful and insignificant location, built into the wall of No 111 Cannon Street, formerly The Overseas Banking Corporation of China. There is a faint possibility that this is merely the top of the original stone, lopped off to make way for today's traffic. Nevertheless, for any part of it to be inconspicuously built into the wall of a bank is a shameful fate for one of London's most important historic relics and it illustrates how easy it is for a materialistic society to destroy our sacred heritage. If restoring it to its original site is now impractical, it should at least be relocated to a more geomantically correct position.

ST.BRIDE'S IN FLEET STREET

Tucked away behind Fleet Street, we have one of London's oldest and most interesting churches. Worship here originated at a 6th century holy well dedicated to St. Brigit or Bride, the White Goddess herself.

The fabric of the current building, designed by Wren, demonstrates very clearly that sacred sites were successively utilised from time immemorial. The crypt of St. Bride's has been excavated (thanks to a WW2 bomb) and shows the remains of Saxon and Norman churches along with some Roman mosaic flooring.

The holy well which was probably associated with a pre-Christian shrine, still survives but is not open to the public. However, the atmosphere in the crypt is definitely worth a visit, particularly a small square Chamber found by following the passageway directly opposite the bottom of the crypt stairs. First impressions tell you the room is small, dark and spooky. If you meditate in there, though, you will find it is actually full of light.

Psychic impressions picked up at St. Bride's included a Saxon Chieftain or local King buried behind the crypt Altar, a 14th Century nun whose cowl appeared to be starched into a pentagonal shape and the surprising image of the Egyptian goddess Isis, with wings of an iridescent blue outstretched with red tips.

TOWER HILL

This is a major junction point of the star's lines. The precise geometric point is not the Tower itself. It is actually a patch of grass on Tower Hill overlooking Tower Gardens and with a pleasant view of the Tower to the South.

It's possible this little vantage point was once a significant ritual site. The Druid Order celebrate the Spring Equinox on Tower Hill every year and have been performing their ceremonies nearby for a considerable number of years. The order's known history dates back to 1717 and they are utilising a spot that has been regarded as exceptionally significant from remote times.

The immediate area is of immense importance, not just for London, but for Britain. The Tower is reputed built upon the Bryn Gwyn, The White Hill, under which Bran's head was buried to act as sacred guardian of these isles. It also boasts physical guardians in the form of the Tower's ravens. Legend warns that the kingdom will fall if the ravens ever leave the tower.

Illustration 32: Horsenden Hill in Greenford.

HORSENDEN HILL

Like Pollard's Hill, this spot in Greenford has awesome views and it seems pathetic to complain that trees prevent you from enjoying them through 360 degrees.

The summit of Horsenden was originally a hill fort but has been excavated, in Victorian times, to allow the construction of underground reservoirs, so any earthworks or obvious archaeological remains have been disturbed or destroyed long ago.

There are one or two earthen banks in evidence, but they're on the golf course which occupies part of the hill and ostensibly they are there to provide obstacles to test golfing skills.

Oddly, the exact geometric point is actually on the golf course, close to a green which has near circular banks built around it, like a mini henge. Of course, there is no telling whether these unusual earth-works were a convenient feature of the landscape when the golf course and greens were created, or whether they were added at that time.

Illustration 33: At Horsenden Hill the goddess apparition resembled another tarot card, Strength.

Psychically, the spot had very powerful presences. By the time I visited Horsenden, I was accustomed to tuning in to the spirit of a place and asking the presence there to manifest in order to convey any information I might need. What surprised me at Horsenden was the first image to present itself was a lion. Nor did it appear to be particularly friendly

The lion was soon joined by a female figure who stood beside it and placed one hand upon its mane as if keeping it under control. She resembled a classical goddess with a long white dress and a garland of flowers in her hair.

The symbolism isn't difficult to interpret. This goddess image is a representation of the Earth Mother in summer. Lions are a solar energy symbol with connections to the element fire, all of which suggests Horsenden was a site with some important solar orientation and possible of ceremonial importance during summer. When asked for a name, more than one came up: Fortuna and Minerva, Mother of the sun.

Once again, this goddess image was reminiscent of one of the Tarot Major Arcana; the card of Strength, so it was becoming clear that the cards were implicated in the mysteries of our sacred sites, perhaps as keys to the archetypal forces that are focused at such places.

Like Pollard's Hill, Horsenden also harboured spirit helpers from the Druid tradition. In this case, two of them, both wearing white robes and headgear that looked rather Egyptian in style. The message they conveyed also had Egyptian connections. They clearly said I should think of Horsenden as a natural pyramid and the hill of Horus. Initially I found this rather puzzling. Horus is the Egyptian Golden Child, the Son of the Sun. He is represented as a falcon or falcon headed god and is associated with the sun at dawn. Why druids should use Egyptian associations, I do not know. Unless perhaps there was some distant unsuspected connection between the two.

Historically, Horsenden Hill is said to derive its name from a Saxon Chieftain called Horsa, whose daughter is thought to be buried beneath Haven Green in nearby Ealing. Horus didn't come into it. However, interpretation of this symbolism as a cryptic message hints, like the goddess imagery, the site has an important association with the summer sunrise. As you will see from later chapters, this proved to be spectacularly correct.

ONE TREE HILL

This place astonished me. It's hidden behind blocks of flats in Alperton Road and approached through a rather dark passageway. To find a pleasant green hill at the other end of it was like journeying through the underworld and re-emerging into the light.

Symbolically, the name One Tree Hill may relate to Saxon mythology rather than the obvious explanation of just a hill with a tree on it. At the centre of Saxon myth is the theme of the World Tree which is the axis of the universe and a natural version of the omphalos stone. Totem poles and Maypoles derive from similar ideas.

In times gone by, a sacred hill with a single tree on it may have been a highly symbolic location, a place of power and the dwelling place of the gods and spirits. Traditionally, the World Tree is an Ash and associated with the Norse God Odin. There are now several trees on the hill, but I didn't notice an Ash.

From One Tree Hill, the view of Horsenden Hill is impressive. Also visible from both is St. Mary's at Harrow-on-the-Hill. These three hilltop sites share an extremely unusual relationship which will be explained later in this book.

Illustration 34: One Tree Hill, Alperton.

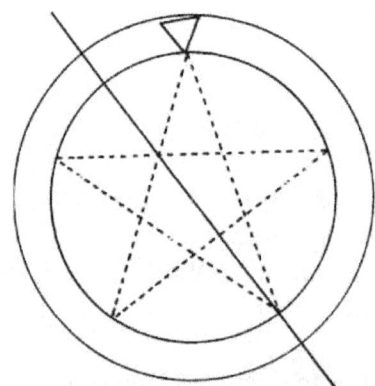

Illustration 35: The Burnt Oak to Bellingham Axis.

THE BURNT OAK TO BELLINGHAM AXIS

This line is not quite so well blessed with sites as the Horsenden axis but with over twenty mark points, it has its fair share and more than a couple of spectacular ones.

1: St. John's, Aldenham, a very ancient church. 2: The Lister Institute, Aldenham Park. 3: A hill in Watling Park, Burnt Oak. This marks the circumference of the circle. 4: Follows the course of the A41 and parallels the A5. 5: Crossroads, Brent Cross A5 and A406 Junction. 6: All Saints Church in Church Walk, Child's Hill. 7: St Andrews at Frognal on the A41, one of the 5-point star junction points. 8: The line runs alongside and parallel to the A41 Finchley Rd from Frognal down to Swiss Cottage. There it directly follows the course of Avenue Rd into Regents Park. 9: Directly through the Inner Circle Regents Park 10: Earthstars centre point near the junction of Hallam Street and Gildea Street. 11: St. Anne's in Dean Street. 12: The National Gallery. 13: Trafalgar Square 14: The Ministry of Defence. 15: The old County Hall on the south bank of the Thames. 16: St. Giles' Church in Camberwell 17: Peckham Rye Park. 18: One Tree Hill at Honour Oak. 19: Beckenham Palace Park. 20: Bromley, a spectacular hill behind the local parish church, topped off by a statue of angels on the local War Memorial. This point is on the outer circle.

THE CAESAR'S CAMP AXIS

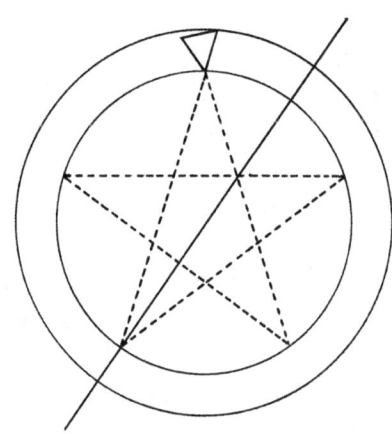

**Illustration 36:
The Caesar's Camp axis**

MARK POINTS

1: A hilltop in the grounds of Chessington Golf Centre.
2: St. Mary's Parish Church at Chessington. 3: The alignment runs directly along a straight section of the A3 Kingston by pass, a distance of at least two miles. 4: A Church at Kingston Rd, New Maiden
5: Wimbledon Common. 6: The Grosvenor Chapel, South Audley Street. 7: Holy Trinity Church, High Cross, Tottenham, a sacred site probably arising from the nearby St. Eloy's holy well.

This line hits very few sites directly, though at least one of them is of exceptional significance: Holy Trinity Church, High Cross, Tottenham. Alfred Watkins remarked on the high incidence of place names on his ley alignments which used the prefix Tot', 'Dod' or Toot'. Fred Fisk's HISTORY OF TOTTENHAM contains many references to the antiquity and possible origins of the town. Although many theories abound, the name, it seems, could evolve from the Druidic/Celtic deity Teutates, equated with Mercury in the Roman Pantheon and Thoth in the Egyptian.
On page 4, we are informed that Tottenham High Cross, which stands on the site of a burial mound; **"was originally Druidical and the name derived from Taute."**

Illustration 37: St. Eloy's Well beside Holy Trinity Church at Tottenham.

Moreover, hilltop sites dedicated to Taut, Toot or Teut are known to have frequently been marked by a stone which literally 'stood' for the god. With the coming of Christianity, such pagan relics were frequently tipped down a nearby well and one such stone was actually found at the bottom of Tottenham's well.

> " **In the bottom when it was attempted to clean it out was found a faire great stone, which had certain characters or letters engraved on it, but it being by the negligence of the workmen broken and sorely defaced, and no man near that regarded such things, it is unknown what they were and what they might signify.**"

Mr Fisk links the dedication of the well to a variant of the sun god Apollo, who was also Helios or Elios, hence Eloi or Eloy. It was certainly a pre-Christian place of sanctity, but for a more detailed understanding of the site's importance, I recommend the fuller account in Mr Fisk's book.

THE SION PARK TO WANSTEAD AXIS

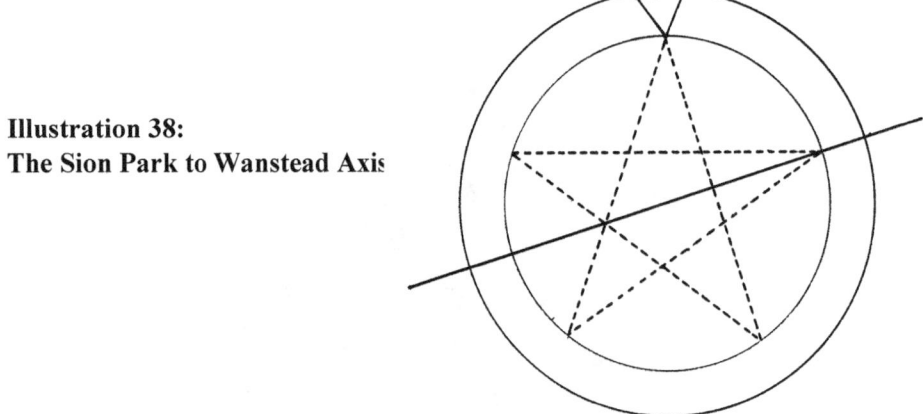

Illustration 38:
The Sion Park to Wanstead Axis

MARK POINTS

1: Valentine Park, Ilford.
2: Wanstead Park
3: St. Gabriel's Church Wanstead Flats
4: Temple Mills
5: The American Church in Tottenham Court Rd
6: All Soul's Langham Place, directly opposite the BBC's central London H.Q. and the nearest church to the Earthstars' centre.
7: The Church of The Annunciation in Seymour Street
8: Kensington Gardens and part of Kensington Palace.
9: Holland Park
10: A church in Fraser St., Chiswick.
11: The Palm House, Kew Gardens
12: Sion Park, site of a fifteenth Century Brigetine Priory.
13: All Saints, the parish church of Isleworth.
14: Church on the edge of Hounslow Heath.

This alignment has far fewer important mark points than the Horsenden Hill axis. The most significant are Sion Park in Brentford, the site of a fourteenth century priory dedicated to none other than Saint Brigit and the parish church at Isleworth which was mentioned in the Doomsday Book. As well as its sites, this alignment is distinguished by the fact that it runs almost parallel to over 16 miles of the A30 from Staines to London and to a large portion of the A12 in Essex.

THE FIVE-POINT STAR INTERSECTIONS

The intersectional lines of the five-point star form a smaller pentagon within the original star. Three of its five mark points are defined by specific structures and they include one of London's most important and most ancient monuments, The Tower of London.

In fact, the exact geometric point here is the natural prominence of Tower Hill, over looking the Tower.

The two unmarked points have indications in their names and locations that they may be concealing more interesting origins which have yet to come to light.

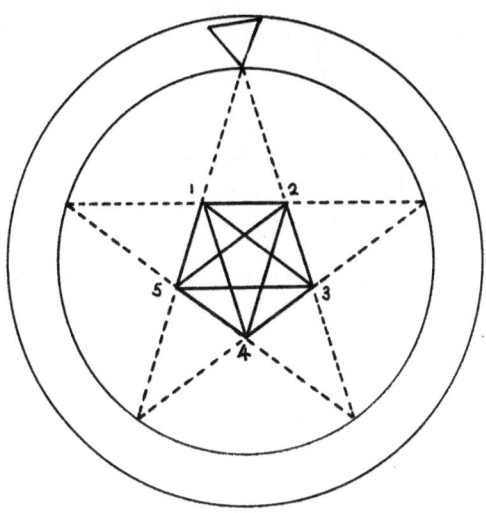

Illustration 39: Inner pentagram intersectional mark points.
I: St. Andrews Church, Frognal at the corner of Frognal and the A41 Finchley Rd in Hampstead. 2: A point on Highbury Hill not far from Arsenal Football Club, regarded as sacred ground by many football fans. 3: Tower Hill, a high point on the grass verge near Tower Gardens overlooking the Tower of London itself. 4: St. George and St. Andrews, Patmore Road, Nine Elms. 5: Point in Abbotsbury Rd, Holland Park.

Illustration 40: The five-point star's intersection point on Tower Hill overlooks the Tower of London and used to be identified by this boundary marker. It has since disappeared.

Illustration 41: Alexandra Palace, one of North London's most familiar hilltops, is directly on the five-point star's line from St. Mary's East Barnet to Bellingham.

Illustration 42:

The Minchenden Oak in the gardens of the former Minchenden/Weld Chapel near Christ Church, Waterfall Rd, Southgate is on the five-point star's line from St. Mary's East Barnet to St. Gabriel's in Wanstead.

SUMMARY

The Earthstars' pentagonal patterns identify a great number of interesting and important places.

The radial axes, on the whole, have a higher score of mark points than the pentagram's other constructional lines. Two of them have far more than their fair share of sites along their length, more than enough to qualify as traditional leys. On the down side, while most of the lines are marked by sites of considerable interest, they would never qualify as leys, in the strictest definition of the term. In fact, a couple of the lines are actually so bereft of notable mark points that this stage left me wondering whether I had found something meaningful, or just a mare's nest of my own creation.

Thankfully, when considered as a whole, the picture is much more encouraging.

The accumulated evidence, particularly the enigma of the parallel roads, seems stacked heavily against the likelihood of this being a totally random freak of nature. A large number of the places identified are truly impressive and include many of the capital's most ancient sacred sites, several of pre-Christian origin, including two holy wells, St. Brides and St. Eloy's at Tottenham; a seemingly overlooked hilltop henge in Norbury; a venerated druidic Oak: Central London's only remaining megalithic monument, the London Stone; a large number of notable high points, including Horsenden Hill, Alexandra Palace, Highgate Hill and Tower Hill, not to mention two "One Tree Hills" and the alignment of Primrose Hill and Parliament Hill to St. Mary's on Church Hill, East Barnet

It's a list which is too long and contains too many extraordinary places to be dismissed as insignificant.

Nevertheless, the shortage of valid mark points on some lines cannot be ignored. At the very least it indicates that what I was dealing with here does not fall within the accepted idea of leys, although it obviously comes under the heading of associated phenomena.

Since most definitions of leys seem to have been arrived at prematurely, before anyone knows for certain what they actually are, this doesn't worry me unduly

What does seem to stand out from the evidence so far is that the phenomenon is very clearly linked to places of natural sanctity, rather than man-made constructions.

The fact that some points are marked by churches does not necessarily mean this pattern was pre-conceived and laid out by human hand. Many of the sites, particularly the hill tops and holy wells, would have been utilised as places of worship from very ancient times when a simple grove of sacred oaks open to the heavens was the equivalent of today's grandest cathedrals and equally uplifting to the spirits.

If I can trace so many spots that could have been our most ancient holy shrines, imagine how many more may be lost and forgotten, built over and buried forever.

Those identified here, and the geometric patterns linking them, are just the tip of the iceberg.

There is much more to this discovery than meets the eye.

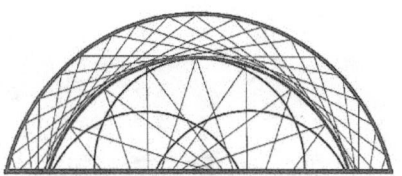

Chapter Six
A Cluster of Earthstars

What did the Barnet Triangle have to do with these latest developments? Ostensibly, an equilateral triangle should have direct connections to a hexagonally based pattern, not a pentagonal one but at first glance, it appeared to bear no logical relationship whatsoever to this five-pointed star.

Having extended the triangle's lines previously without much success, I decided to re-examine them in the context of the new patterns. My assumption that they were unrelated proved to be totally incorrect.

Firstly, I extended the western side of the triangle, the line from St. Mary's Monken Hadley to St. Mary's East Barnet.

I was surprised to find that it meets the circumference of the circle at exactly the same point as one of the pentagonal axes; A spot in New Charlton diametrically opposite Horsenden Hill (see illustrations 43, 44 and 45).

Unfortunately, this isn't a definitively marked site. There's an old St. Catherine's church nearby, or at least there was when I visited the area some years ago, but nothing particularly remarkable on the exactly geometric point. Nevertheless, the connection between the Barnet Triangle and the five-pointed star was clear and direct. More importantly, the resultant chord is not any old line across a circle. It marks a precise division of the circumference in the ratio 2:3.

Encouraged by this, I measured the angle between the chord and the diameter. Then, by reflecting the line off the circumference at the same angulation, I continued it around the circle. I don't know what prompted me to do this. Call it intuition. It just felt like the right thing to do at the time. The result was startling.

The reflected line struck ten evenly spaced points around the circle, creating a ten-pointed star whose mark points include St. Mary's East Barnet, Pollard's Hill, all five points of the original pentagram and all five of their diametric opposites.

This is a fascinating development after the five-point star because a single pentagram is not a balanced geometric figure. Its five axes identify ten points on the circumference of a circle, so its balanced form is actually

Illustration 43: An extension of this Barnet Triangle line identifies a point on the inner circle diametrically opposite Horsenden Hill.

two interlocking pentagrams, or a single ten-point star like this one. Consequently, a ten-point star is a perfectly natural progression from the pentagram. To find it was also a natural development from the Barnet Triangle was absolutely amazing.

This confirmed beyond a shadow of a doubt that the two were not separate geometric figures. They were connected and it was definitely no coincidence.

Amongst the new sites identified by this figure are Sion Park in Brentford and Watling Park in Burnt Oak. Both are places of considerable interest as ancient sacred sites.

TEN-POINT STAR MARK POINTS CLOCKWISE:

1: St. Mary's East Barnet.
2: Point in Lee Valley Park, Enfield.
3: St. Gabriel's, Wanstead Flats.
4: St.. Catherine's New Charlton.
5: Bellingham Green.
6: Pollard's Hill, Norbury.
7: Caesar's Camp, Wimbledon.
8: Sion Park, Brentford
9: Horsenden Hill, Greenford
10: Watling Park, Burnt Oak

London City of Revelation

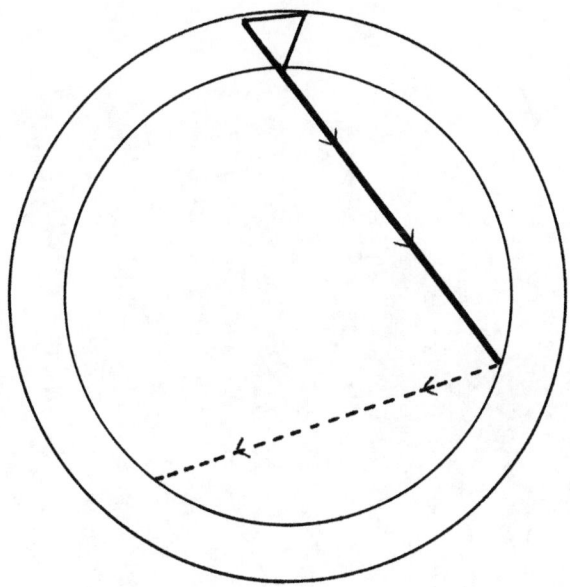

Illustrations 44 & 45: Extending the western side of the Barnet Triangle into the circle creates a perfect ten-point star which incorporates all five points of the original pentagram.

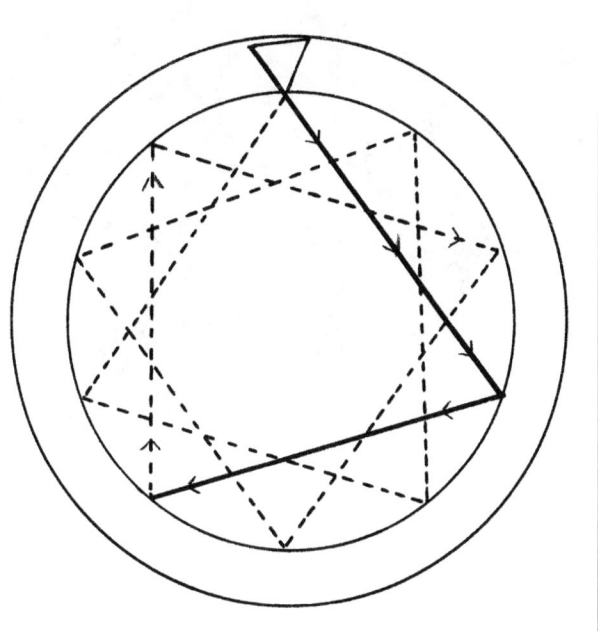

London   City of Revelation

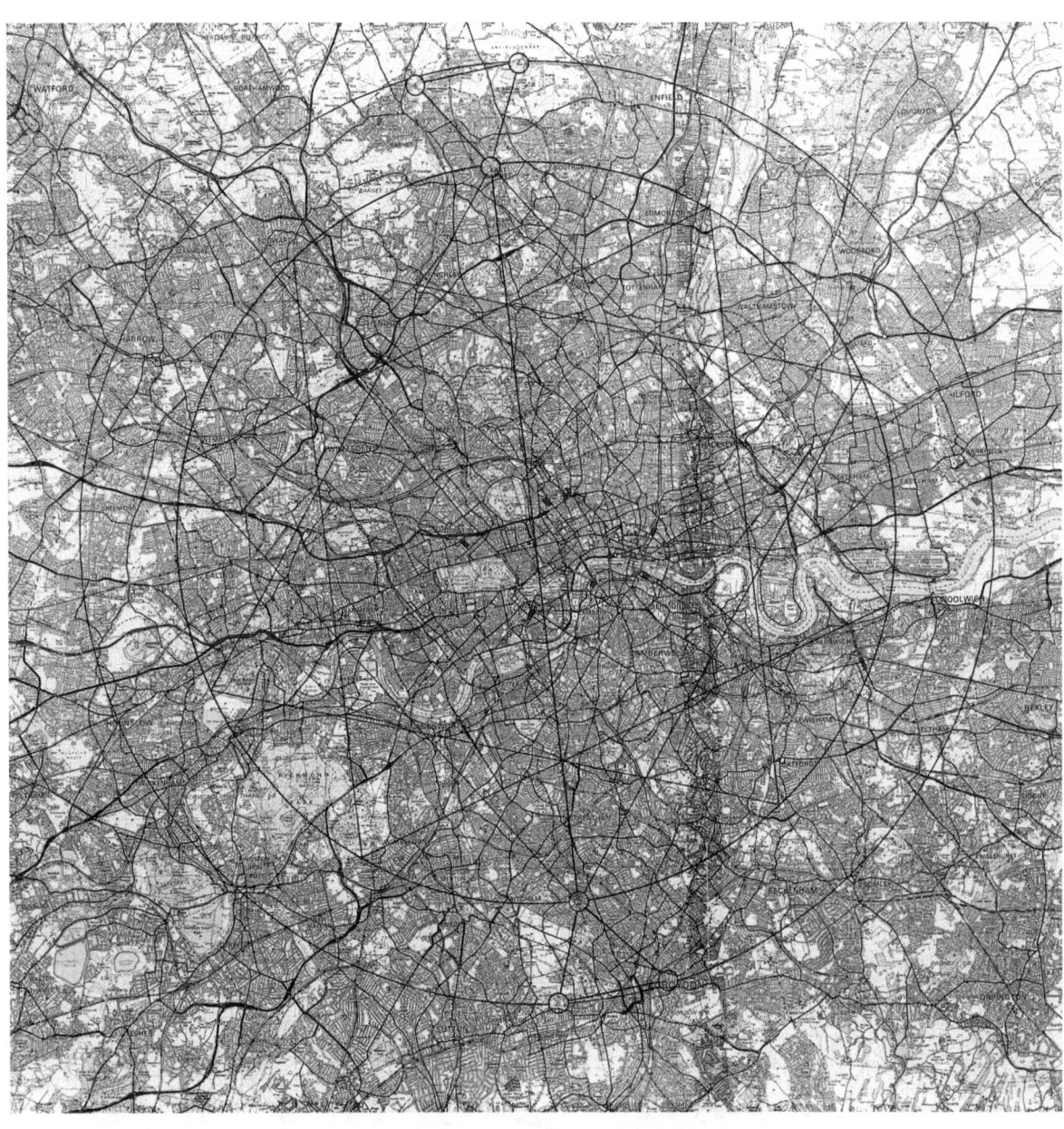

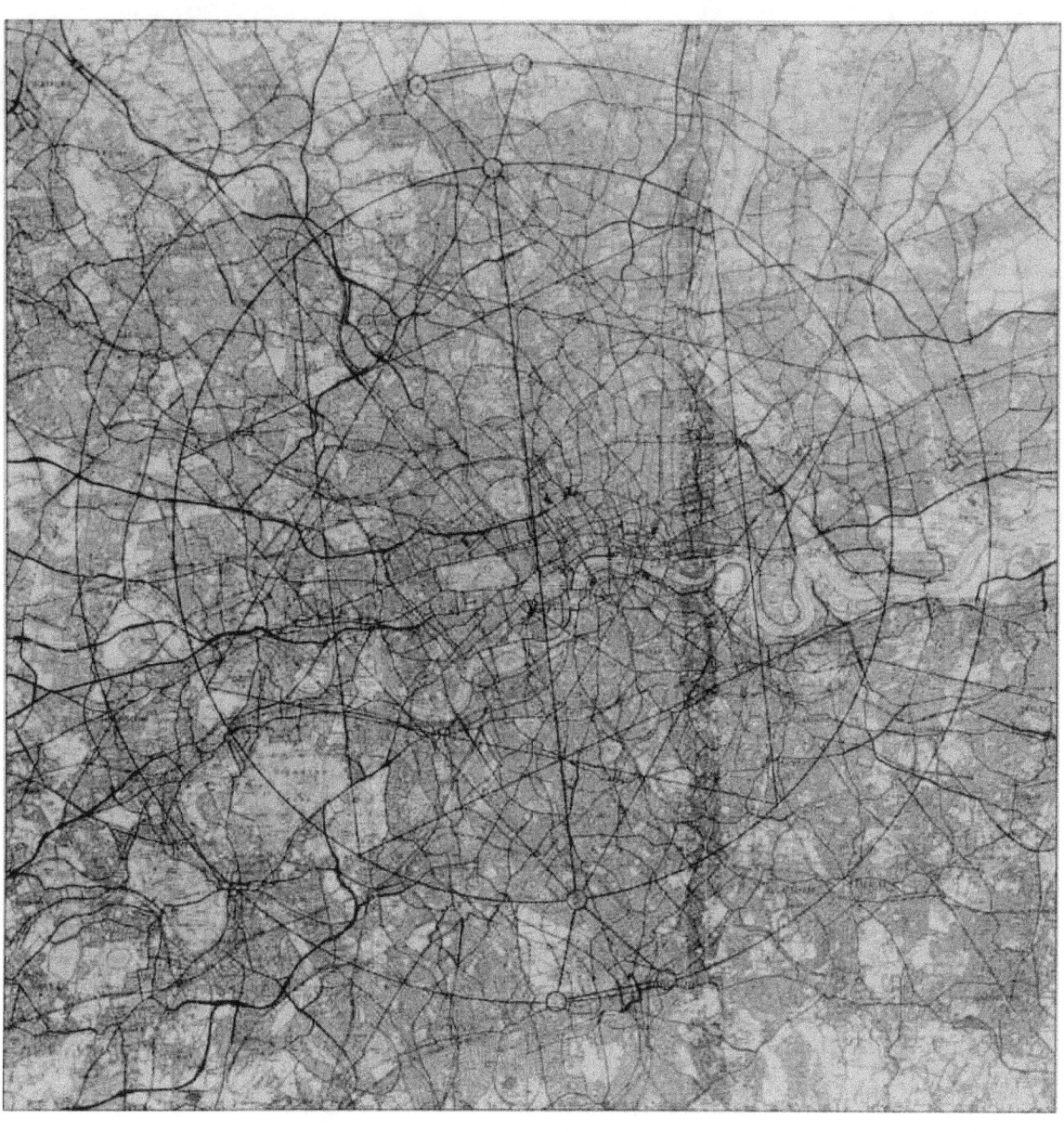

Illustration 46: The 10-point star generated by extending the Barnet Triangle's geometry into the Earthstars circle. O.S. map © Crown Copyright. All rights reserved. License No. 100029558.

SION PARK

Sion Park was the site of the Monastery of Sion, founded by Henry V in 1415 for the accommodation of sixty nuns and twenty five monks belonging to the Order of St. Bridget. As we have seen in previous chapters, sites dedicated to Saint Brigit may have associations with the White Goddess and like Saint Bride's in Fleet Street, this house of God may have been established on or near a pre-Christian shrine. The site of the Monastery does not coincide with the exact geometric point in Sion Park, but a large mound of indeterminable age does. Unfortunately, it's in private gardens not accessible to the general public, but it can be clearly seen from inside Sion House, which is open to visitors.

WATLING PARK BURNT OAK

Burnt Oak is an evocative name suggesting a venerated tree and Druidic associations. The precise geometric point here is in Watling Park which contains an inspiring hill. Since hilltops were frequently "places of the gods" to our ancestors, this may be significant.

The hill's crest certainly had a pleasant atmosphere and although there weren't quite enough oaks to be a grove these days, I felt it had been in the distant past. On a more recent visit, in 1997, I noticed a strange circle of discoloured grass near the top of the hill, (illustration 47). It was about twelve feet in diameter and quite distinct.

There were no other marks either around the area or within the circle and no obvious flattening of the grass as there is with crop circles.

THE EIGHT-POINT STAR.

Having found a new and remarkable addition to the geometry constructed from one side of the Barnet Triangle, the next step followed logically: to see what might evolve from extending the other two sides.

The eastern side of the triangle turned out to generate an equally impressive result. Extending the line from Camlet Moat through St. Mary's East Barnet, then across the circle, produced another star pattern.

The line touches the inner circle circumference at Kingston Vale on the south side of Richmond Park. The site is not particularly remarkable in itself, but the geometry it reveals is. It is a point that would divide the

Illustrations 47: The unusual grass circle on the hill in Watling Park, Burnt Oak.

Illustrations 48 and 49: Extending the Barnet Triangle's line from Camlet Moat to St. Mary's East Barnet generates a perfect eight-point star within the circle.

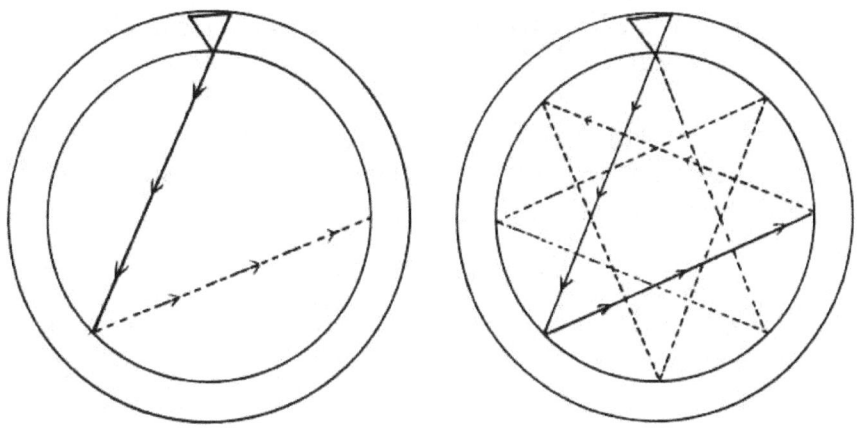

circle's circumference in the exact ratio of 5:3 and when the line is reflected around the circle in the way I had done with the ten-pointed star, it identifies eight equally spaced divisions of the circumference, creating another perfect star pattern: an eight-pointed star. The pattern is aligned to the original north-south axis, so St. Mary's East Barnet and Pollard's Hill in Norbury both feature prominently as the north and south points of the star respectively.

On the modern O.S. maps nothing special seems to mark either the east or west points and local visits seemed to confirm they were undefined. Or so I thought until Margaret Wood, who lives in Manor Park, wrote informing me that she had unearthed some fascinating facts about both places.

Prior to the modern building development in the area, the East Ham point had been market gardens. Amongst them existed a number of ponds, including a well with healing waters. THE VICTORIAN HISTORY OF ESSEX, Vol. VI refers to it as ;

"Miller's Well, a Medicinal spring situated at the point where the present Cheltenham Gardens joins Central Park Road."

A book on East Ham Village, EAST HAM, FROM VILLAGE TO COUNTY BOROUGH by Alfred Stokes, one of its former Mayors, adds to the information:

" One of the most interesting spots in the village was The Miller's Well, situated at the junction of what are now known as Rancliffe and Central Park Roads. Its water was famous for miles around as it was said to possess peculiar properties which were supposed to help forward the cure of many ailments. In particular it was reputed to be very beneficial in complaints affecting the eyes."

A friend of Margaret's, Joan, had lived in the area since childhood and recalled that a shop which had stood on the corner of Central Park Road had been plagued by subsidence because of problems with underground water. An 1898 O.S. map of the area showed both a spring and a well in the immediate vicinity of the Earthstars circle eastern point. Margaret suggested that, in prehistoric and/or Roman times, any such

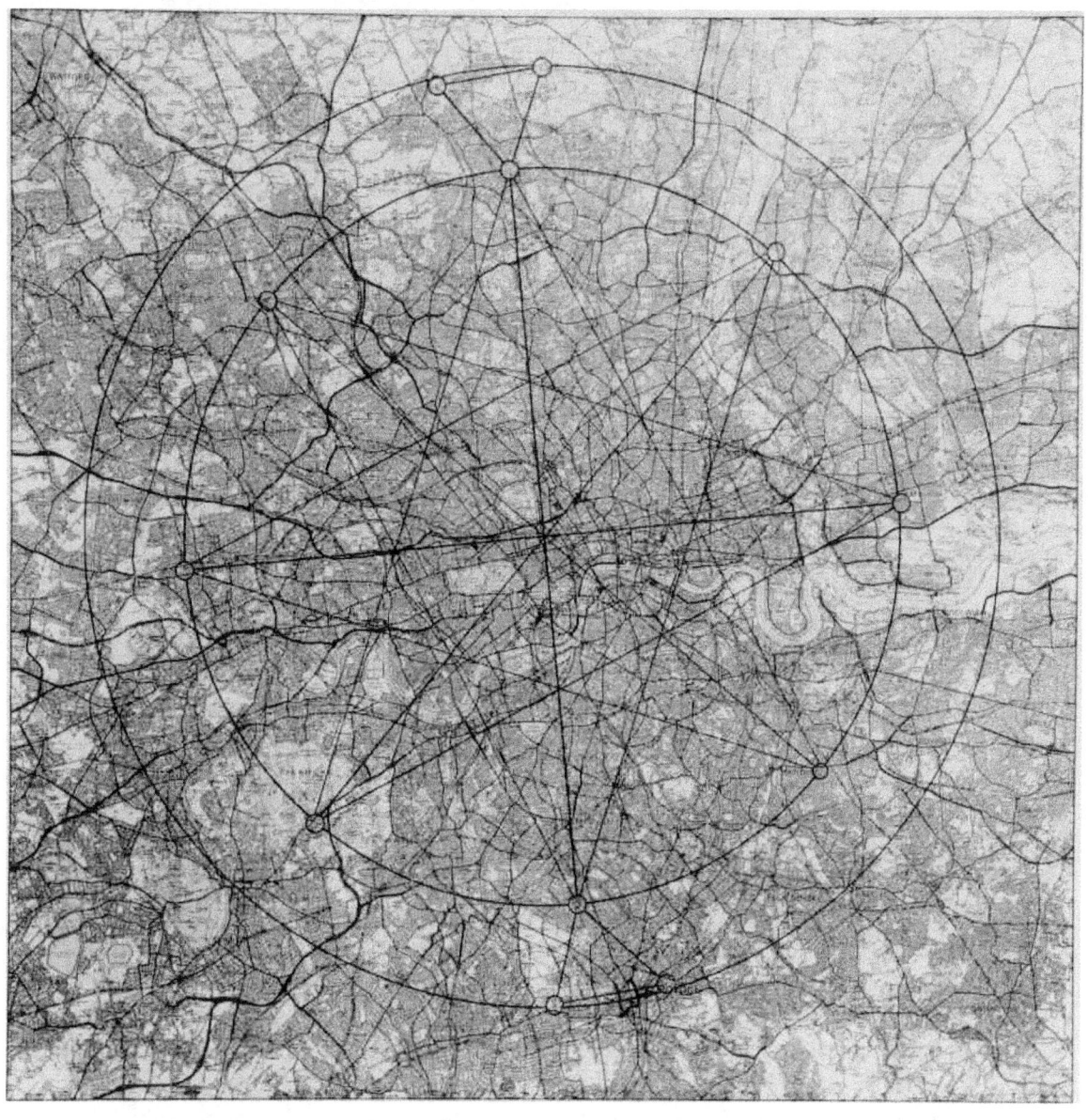

Illustration 50: The eight-point star generated by the eastern side of the Barnet Triangle. Mark points clockwise; 1: St. Mary's East Barnet. 2: Playing fields near Chingford Rd, Enfield. 3: The Miller's Well (now lost) near Central Park, East Ham. 4: Hither Green Methodist Church. 5: Pollard's Hill Norbury. 6: A point near Ulleswater Crescent Kingston Vale. 7: Western point of the circle (former site of a markstone) near Kensington Cemetery, Hanwell. 8: Station Parade, Queensbury (point on the grass central reservation). O.S. map © Crown Copyright. All rights reserved. License No. 100029558.

medicinal spring or well would have given rise to a nearby shrine, probably dedicated to the deity of the waters.

By coincidence, Margaret also had contacts in the area where the circle's western point lay, so she then turned her attention in that direction. On an out-of-date Ordnance Survey map, she found a stone marked on the south side of Uxbridge Road, a few yards west of Grosvenor Road. The boundary between Hanwell and Ealing runs nearby and as Boundary Cottage was clearly marked on the map, she suggested that this may originally have been a boundary stone. So it seems the Earthstars circle's east and west points are marked almost as significantly as the north and south.

Like the pentagonal star, the eight-pointed star's constructional lines identify a number of important sites.

The hilltop in Finsbury Park features prominently in this star pattern, on a line which includes a number of other hills whose significance may have passed unnoticed until now.

From St. Mary's Church Hill in East Barnet it goes directly to a hill currently occupied by Brunswick Park Cemetery, then onto another prominent hill at New Southgate, marked principally by a large traffic island at the junction of Brunswick Park Rd, Waterfall Rd and Friern Barnet Road. The next hill top is a piece of private woodland used by scouts at Woodfield Way in Bounds Green. Then, the spectacular hilltop in Finsbury Park. The remaining mark points on this line are; Green Lanes Methodist Church in Stoke Newington and a straight section of Green Lanes to the south of it, The R.C. Church of St. Mary and St. Joseph in Balls Pond Rd, St. Paul's Parish Church in Shadwell and St. Saviour's R.C. Church in Lewisham High Street.

Until I traced this alignment, I had no idea that there was an impressive hill in Finsbury Park. One visit though was enough to confirm that it is a powerful spot from the ley and 'Earth energy' perspective. From the top of the hill there are clear sighting lines to other prominences which could form the basis of ley alignments. One to Highgate Hill, for instance, extends directly to the Ordnance Survey triangulation point on Barn Hill in Wembley and then on to St, Mary's Church at Harrow on the Hill. Moreover there is some landscaping to the west of the pond which resembles the remnants of a henge-like earthworks.

That's not all. Walking up the hill, I couldn't help but notice that there are number of small, odd-looking stones grouped together on the south east side of the hill. They're scattered about haphazardly, but I suppose they could be someone's idea of a rockery or landscaping project.

Illustrations 51a and b: Some of the stones on the hillside in Finsbury Park (Shouldn't someone make them into a nice circle ?)

It occurred to me that if a megalithic monument had once crowned the hill, these broken fragments could be re-arranged to create a new one. Coincidentally, the next hill along the same alignment is also strewn with boulders, much larger than those in Finsbury Park. Most lie on the hillside to the west of Durnsford Road, Bounds Green and are clearly visible. Here they do look like part of a landscaping project but they could equally be the remains of a megalithic monument which has been cleared from a site nearer the summit of the hill where a modern housing development now exists.

THE 8-POINT STAR'S NE TO SW AXIS

There aren't many earthwork encampments around London. Caesar's Camp on Wimbledon Common is one and the NE-SW axis of this eight-point star hits three more; Amesbury Banks and Loughton Camp in Epping Forest, plus the possible site of a third camp, known as The Brill which the antiquarian William Stukeley believed to have been in the area of the Somers Town Estate between Euston and St Pancras' old church.

This same alignment also passes through the impressive St. Pancras church opposite Euston Station which, coincidentally is topped by an octagonal tower based on the ancient Tower of the Eight Winds in Athens. A little further along Euston Rd, the alignment connects to a less impressive concrete edifice, the office block which housed CDP Ltd, an advertising agency where I worked at the time I made this discovery The offices have since been transformed into a posh hotel.

Other mark points on the line are; The Cathedral Church of the Good Shepherd, in Rookwood Rd, N16, near the crown of Stamford Hill (its sign proclaims it as an ancient Catholic Church);The Ancient Mother Church of Stoke Newington at Clissold Park (opposite the more recent St. Mary's); St. Pancras Church, Euston Rd; All Souls, Langham Place (the closest church to the centre of the Earthstars patterns); St. Peters in de Vere Street (strange it should go through All Soul's and St. Peter's as they are linked historically); Hyde Park; The Victoria and Albert Museum; St. Mary The Boltons; a point behind West Brompton Cemetery next to the Chelsea ground; All Saint's Church, Fulham on the north side of Putney Bridge; a high point in Richmond Park near the Broomfield Plantation.

Mark points on the NW-SE axis are less numerous; a reservoir near Gondar Gardens in West Hampstead; Regents Park, where the line runs parallel to a path (strangely, the paths in this part of the park are laid out in an irregular five pointed star pattern); Cambridge Circus W1; Waterloo

Illustration 52: All Soul's Langham Place, the nearest church to the Earthstars centre. The eight-point star's NE-SW axis runs down the centre of the building.

Bridge traffic island; St. George's Church in Burgess Park.

On the East-West line are: the site of the Miller's well in East Ham; St. Bartholomew's (now Steeple Court) in Coventry Rd, Bethnal Green; St. Luke's Church in Fann Street; The British Museum; The Catholic Church of Our Lady of The Rosary in Marylebone Rd; All Saint's Church in Colville Gardens, Notting Hill; St. Aidan's R.C. Church in East Acton and the site of the boundary stone in West Ealing. The most startling discovery on this alignment though is the fact that it can be extended directly to Silbury Hill in Wiltshire, Europe's largest man-made mound and part of the Avebury sacred landscape. Extended further to the west, it reaches the Bristol Channel near Brean Down and can be continued into the channel to Lundy Island. To the east, it passes through Prittlewell Priory in Southend.

The N-S axis of the eight-point star is shared with the 5 and 10-point patterns, so its numerous mark points have been listed in earlier chapters. There is a definite connection between the eight-point star and the five/ten-point stars and not just through their common origins in the Barnet Triangle and its associated circles. Each star contains within its geometry identical replicas of itself at a reduced scale.

Illustration 53: Beyond London, the main East-West line goes directly to Silbury Hill via several other prominent mark points and reaches The Bristol Channel near Brean Down.

Illustration 54: The pentagram and eight-point star share an inner circle through their intersectional points.

If the two Earthstars patterns are overlaid, it becomes obvious that their inner geometry interconnects, most noticeably via a shared inner circle. A circle drawn through the five intersections of the pentagonal star just happens to sit perfectly inside the inner octagon created by the eight-point star lines (Illustration 54).

If you recall, the inspiration to look for these two figures came together, hinting that there was some possible association between the two. It is remarkable to discover that they interconnect in ways I had never imagined.

THE CAMELOT PENTAGRAM

My attempts to extend the Barnet Triangle's sides into larger constructions didn't just uncover one more Earthstar. The same line which gave me the eight-pointed star could also be extended further, to the circumference of the outer circle where it formed the basis of another perfect star pattern.

By following the angulation of the line around the circle in the same way as before, it formed a pentagram even bigger than the first, with Camlet Moat as its upper point. Once again, the star had remarkable connections to the original geometry. The constructional lines of this pentagram interact with all five points of the original one. This makes a connection between the outer circle and the inner one, so it may be no coincidence that this pentagram is involved in a further relationship between the two circles.

When its five points are linked as a pentagon on the outer circle, it appears to hold the inner circle within it (Illus 57).

As if that were not enough to be convincing, this figure also shares a relationship with the Croydon triangle geometry. You merely have to draw in the axes of the star for this to become apparent. St Mary's at Beddington proves to be diametrically opposite Camlet Moat. This implies a decagonal figure, so if the geometry was progressed to create a ten-point star from the five axes, St. Mary's Beddington would be the southern-most point.

This central axis between Camlet Moat and St. Mary's at Beddington is aligned virtually north - south, whereas the central axis of the original pentagram, ten and eight point-stars is a few degrees off but may correspond to a specific magnetic north alignment.

Illustration 55

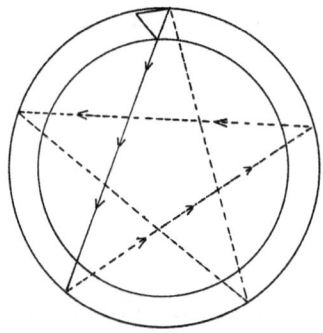

Illustration 56

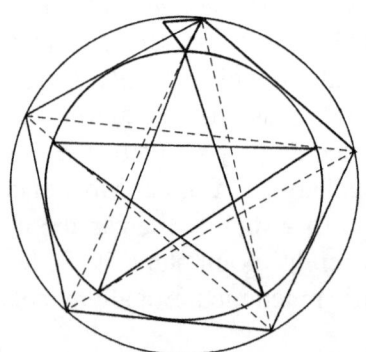

Illustration 57

Illustrations 55 & 56: Extending the eastern side of the Barnet Triangle creates a larger pentagonal figure on the outer circle. Its constructional lines intersect all five points of the original star.

Illustration 57: As a pentagon, it encloses the original circle.

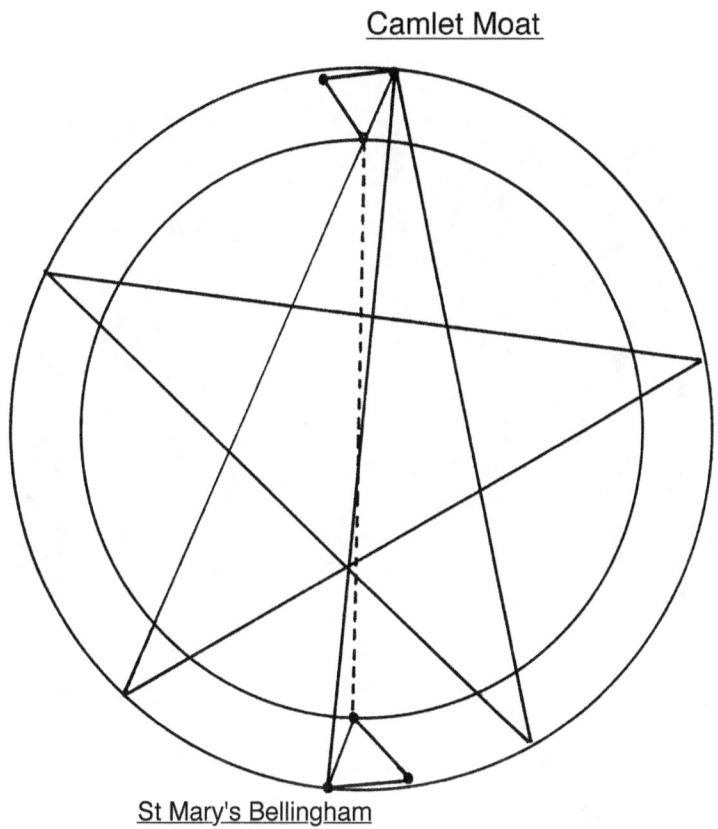

Illustration 58: Diagram of the relationship between Camlet Moat and St. Mary's Beddington. They are diagonally opposite which implies another ten-point star rather than a single pentagram.

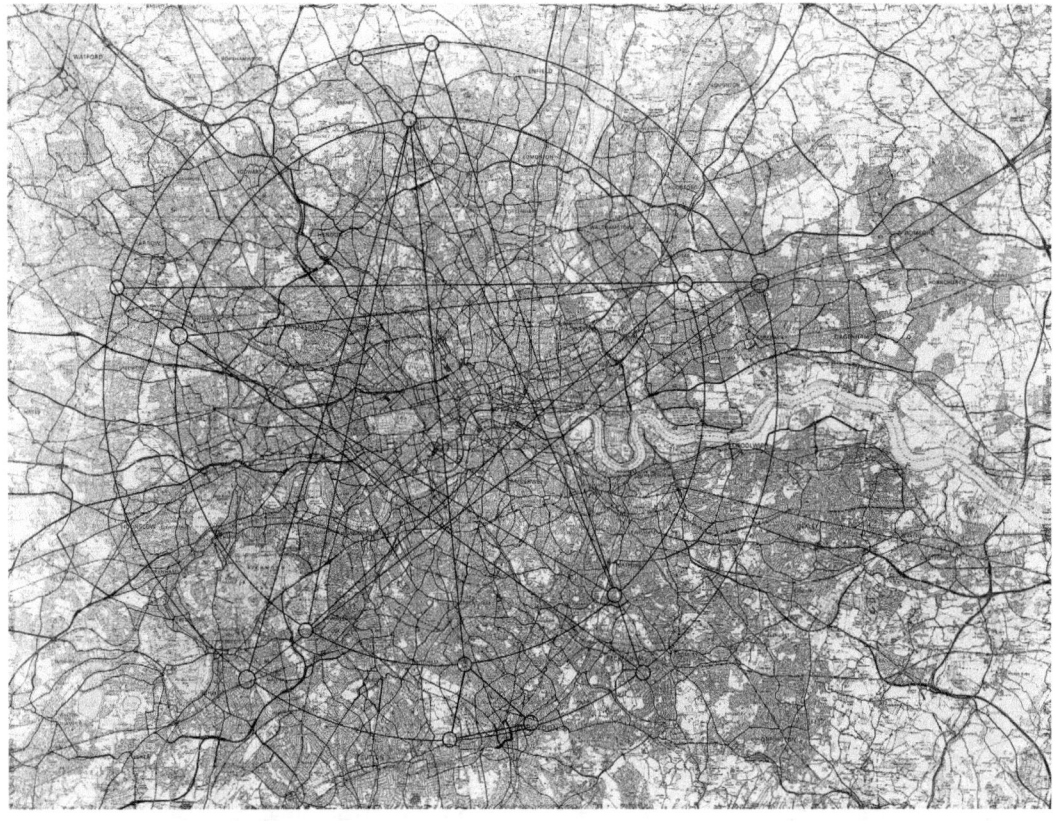

Illustration 59: Map of the Camelot pentagram. O.S. map © Crown Copyright. All rights reserved. License No. 100029558.

THE EIGHTEEN-SIDED POLYHEDRON

So far, two sides of the Barnet Triangle had produced a spectacular series of inter-related patterns. What would the third side reveal ?

The geometry was a little more difficult to construct but extending the line from Camlet Moat through St. Mary's Monken Hadley creates an angle of 80 degrees to a radius of the larger circle, so the line can be reflected around its circumference at an angle of 160 degrees. This creates a totally different design to the star patterns previously generated by the method, a perfect 18-sided polyhedron.

Halfway around the circumference, I discovered something else. I could have created the exactly same pattern from the Croydon Triangle. The line from St. Mary's Beddington to St. Mary's Addiscombe is part of the same figure. It would have generated exactly the same pattern.

This extraordinary geometric relationship links the Barnet and Croydon triangles across twenty miles of suburban London and it is not the only connection this pattern reveals.

The large pentagram's axis from Camlet Moat to St. Mary's Beddington is a shared feature of this polyhedron.

Once again, the geometry proves not to be an isolated figure but one that is closely inter-connected with the existing patterns.

In fact, the Barnet and Croydon triangles are directly related by a surprising number of remarkable geometric relationships, despite the fact that they are on opposite sides of the capital, have no other obvious links and were initially discovered by an extremely unusual process.

As if it were needed for a final, conclusive piece of evidence, there is one more pattern which proves beyond a shadow of a doubt that these two triangles and all the other geometric figures are linked.

Another five-pointed star.

MARK POINTS OF THE 18 –SIDED POLYHEDRON
1: Camlet Moat.
2: A point in playing fields in Enfield near the Queen Elizabeth stadium.
3: A hill crest in Station Rd, Chingford.
4: A small wood bordering parkland in Heathcote Avenue, Clay Hill.
5: A point in Highfields Rd at Woodford Green.
6: Recreation ground near Plumstead Marshes.
7: Grounds of Charlton R.F.C. behind the Parish Church of The Holy Trinity, Eltham.
8: St. Martin's Hill Park, Bromley.
9: Catholic Church of Our Lady of The Annunciation, Bingham Rd, Addiscombe.
10: St. Mary the Virgin, Beddington.
11: St. Anthony's Hospital on the A24.
12: The Chapel of Kingston Cemetery, Norbiton.
13: A point very close to All Saints Church, Twickenham.
14: Central grass reservation in The Glen, Norwood Green at the Crane water Park end.
15: Northolt Methodist church near junction of A40 and A312 (ancient crossroads).
16: Corner of Harrow View and Headstone Gardens.
17: A point in fields near A41 close to Elstree. On the O.S. map a spring is marked in the vicinity.
18: Open field to the west of Galley Lane in Arkley.

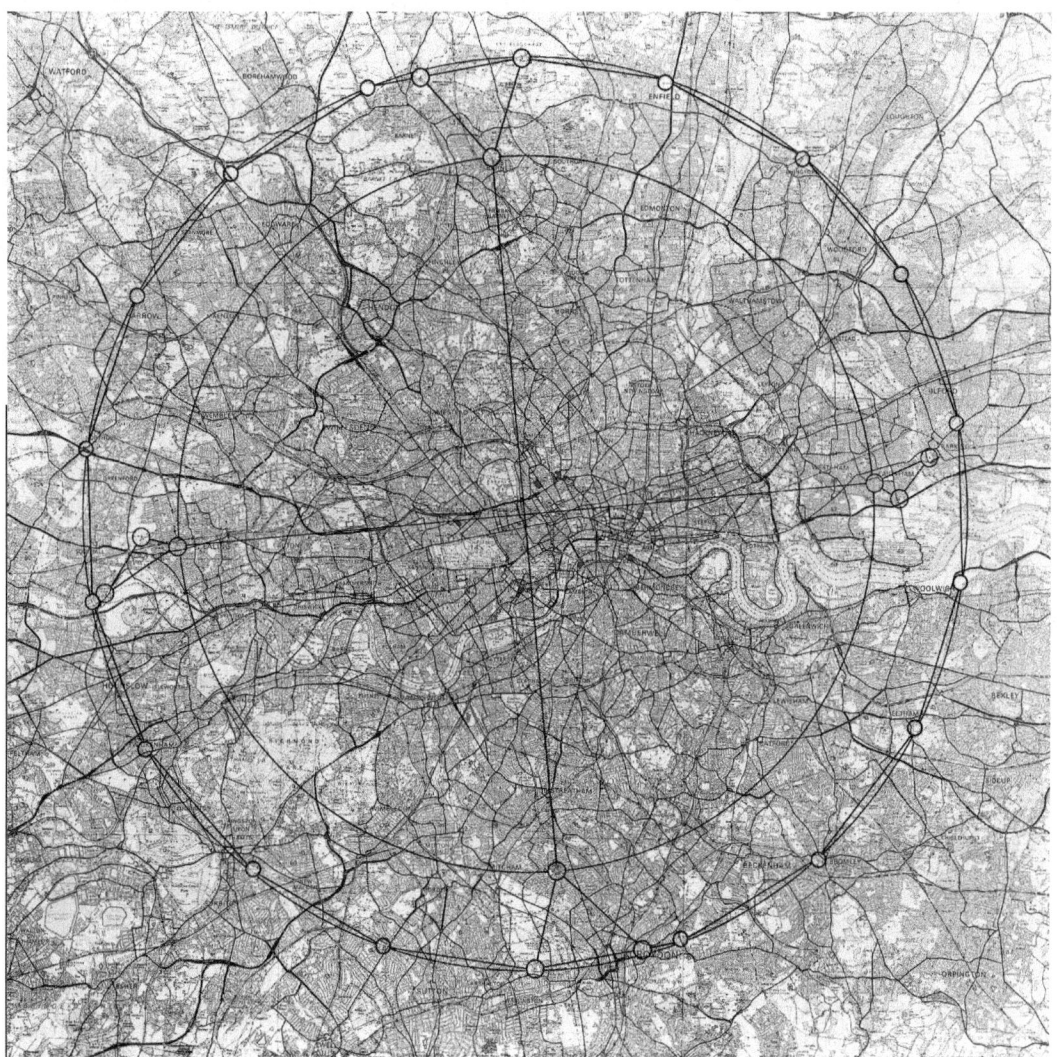

Illustration 60: The eighteen-sided polyhedron which can be generated from the outer side of either the Barnet Triangle or the Croydon Triangle. O.S. map © Crown Copyright. All rights reserved. License No. 100029558.

THE CROYDON TRIANGLE'S STAR

One side of the Croydon Triangle can be extended to create a further pentagonal star; the line from St. Mary's, Beddington to Pollard's Hill. When traced across London, it meets the inner circle circumference in Lea Valley Park, Enfield. This happens to be an important spot already linked to the original five-pointed star. It is diametrically opposite its Caesar's Camp point and so marks the end of that particular pentagonal axis. It is also one of the ten-point star's defining sites.

As if to emphasise that this is not by chance, when progressed around the circle by my usual method, it strikes all five of the original Earthstars' pentagram's diametrically opposed points to create another five-point star, a mirror image of the first.

Together they form a perfectly matched pair, sharing all their points in common with the ten-point pattern. All three patterns are totally inter-related.

More important perhaps, all the other patterns are equally interrelated as well. The Barnet Triangle, for instance is connected directly to the construction of the five, eight, ten and eighteen-point patterns, all of which interconnect with the other figures in some way.

All are obviously part of the same overall design.

The discovery cannot possibly be the result of chance.

There too many coincidences to be a coincidence.

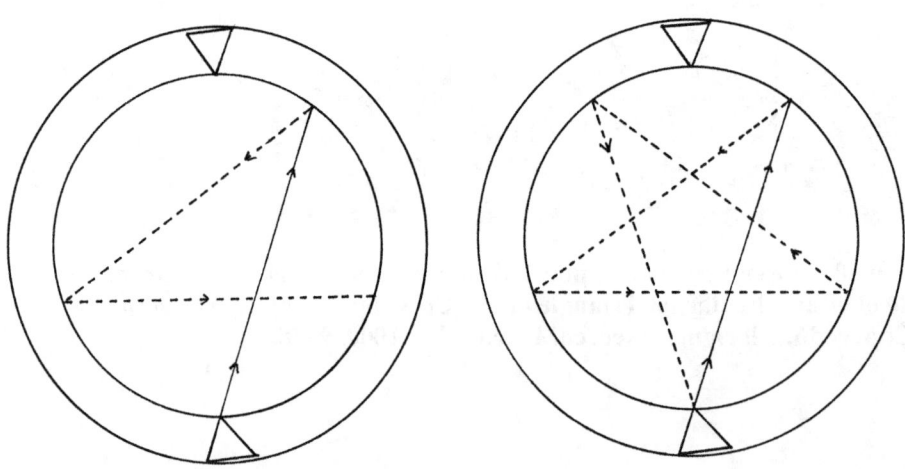

London   City of Revelation

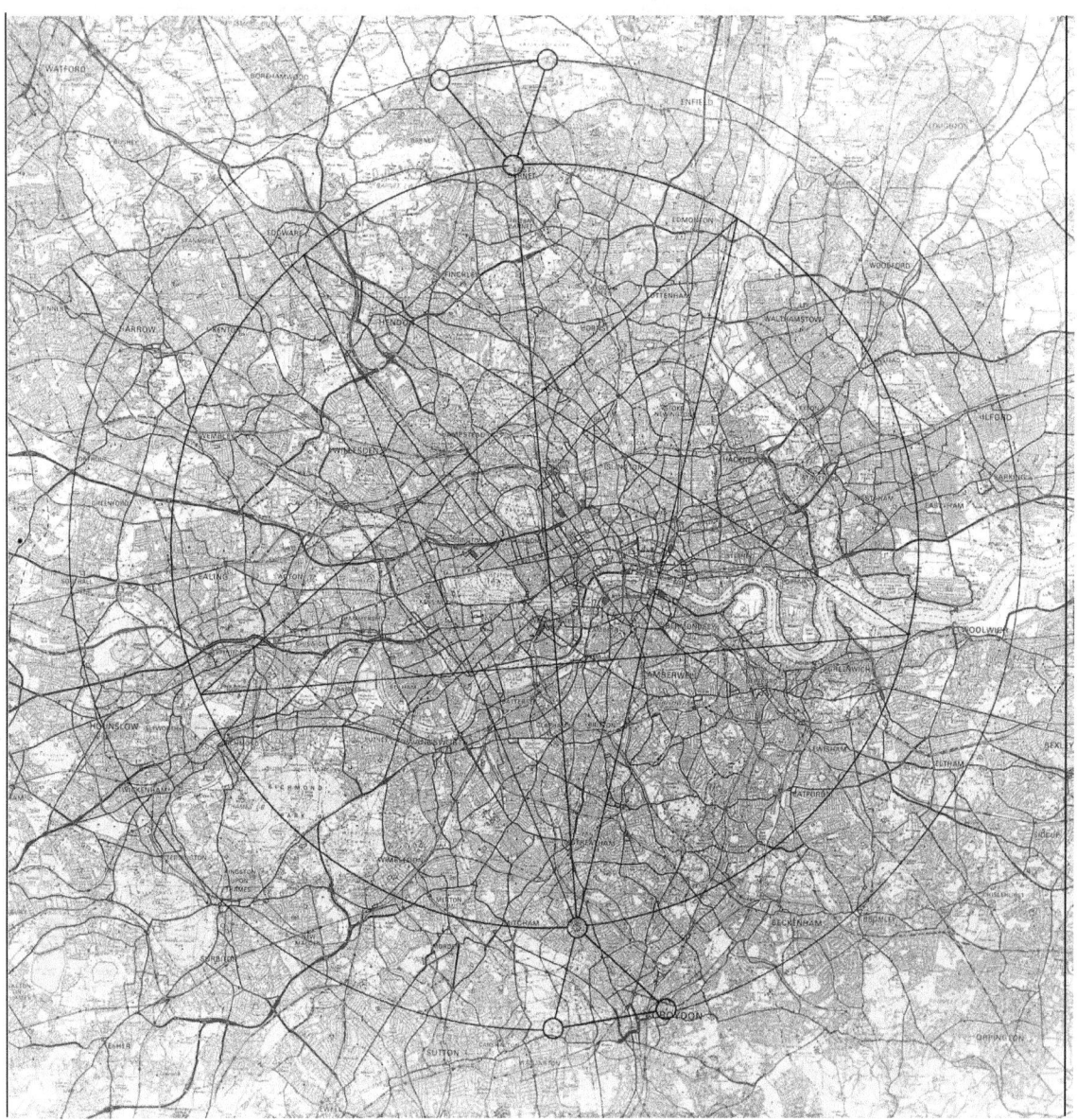

**Illustration 61 & 62 opposite: The construction of the Croydon Triangle pentagram.
Illustration 63 above: The Croydon Pentagram map. Mark points previously listed as part of the 10-point star. O.S. map © Crown Copyright. All rights reserved. License No. 100029558.**

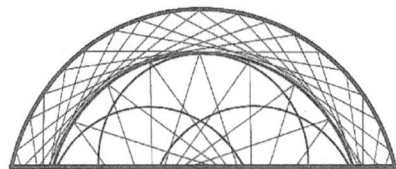

Chapter Seven
The Divine Plan

In some respects, it is hardly surprising that geometry looms large on the landscape. A close inspection of the world at large reveals an enormous amount of recognisable geometric structures. Wherever life's rich pattern is clearly discernible, the same few basic geometric building blocks, which include pentagonal and hexagonal forms, consistently recur as a blueprint for all manner of physical structures, animal, vegetable and mineral.

They can be seen in the arrangement of petals on flowers, the arms of a starfish, in the shape of snowflakes, honeycombs and quartz crystals, to name but a few.

A more discrete pattern runs through creation in the form of Plato's Golden Section, a holistic law of proportion that ensures every part of a structure shares a perfect and harmonious relationship with every other part as well as with their sum, the whole. In a holistic universe, this should not come as a surprise. Amongst the structures which demonstrate Golden Mean proportions are snails' shells, spiral galactic nebulae and a large number of things in between, including human bodies. Mathematically, the Golden Mean cannot be calculated absolutely. Its closest approximation is a ratio of 1:1.618033 ... The mathematician Filius Bonaccio (better known as Leonardo of Pisa or Leonardo Fibonacci) discovered that it can also be expressed as a sequence of numbers where each one is the sum of its two predecessors. For example; 1,2,3,5,8,13,21,34,55,89,144,235, and so on. This is commonly known as a Fibonacci series.

The Golden Mean, however, is derived primarily from a geometrical relationship rather than a numerical one and interestingly, the figure that best expresses it is none other than the pentagram, whose intersecting lines divide each other in perfect Golden Mean proportions, 1;1.618.

From this perspective, the pentagram is a perfect expression of nature's laws of proportion which elsewhere result in a great many naturally beautiful forms. Vitruvius' well known illustration of a man superimposed upon a pentagonal star is not a superficial analogy. It demonstrates that both are based upon the same mathematical laws of proportion.

Illustrations 64, 65, 66 & 67: Some of nature's pentagonal and hexagonal structures; pentagonal symmetry common to many flowers; a starfish; the hexagonal structure of a snowflake; a micro-electrograph showing the hexagonal formation of uranium atoms.

The pentagram's embodiment of the Golden Mean meant it was held in high regard by our scholarly ancestors. Pythagorus chose it as the symbol for his School in Crotona which he founded in 500 BC and to him, the five-point star was a token of health and life. In mediaeval times, it was used as a symbol of protection to ward off evil. The observation that a few geometric forms constantly recur throughout the diversity of structures in the natural world may be taken as a clear indication that the forces of creation work along strictly geometric lines and that underlying them are recurring numerical and geometric relationships.

In FERMAT'S LAST THEOREM, Simon Singh says:

"Pythagoras realised that number was hidden in everything from the harmonies of music to the orbits of planets and this led him to proclaim that; everything is number."

Another much quoted tenet of the Pythagorean school is: **"God is a geometer."**

This may be regarded as an over simplification by some, nevertheless, as a general principle, the notion has given rise to an area of study known as sacred geometry which seems to have existed in various forms, shrouded in varying degrees of secrecy, for several thousand years.

SACRED GEOMETRY

Since the Earthstars patterns are sacred geometry, as distinct from any more mundane kind, an understanding of it should provide some vital keys to this mystery

Sacred geometry is founded on a simple, but far-reaching premise. It is a study of how the forces of creation shape the world; an investigation into the very structure of the universe.

From this perspective, geometry must have held immense importance for those who understood this symbolism.

Every point was understood to encapsulate the unlimited potential of the universe's creative energy. Every line was a line of divine force.

Every angulation, a modification of that force affecting its nature and purpose, just as light is modified and broken down into its component elements by the effect of its angulation through a prism.

Every geometric pattern was a symbolic circuit diagram of how the life-force of the universe functions in the material world.

Geometry gave form and definition to the intangible and infinite, enabling those who could appreciate its inherent symbolism, to gain an insight into the hidden forces at work within nature and to discern, to some degree at least, the divine plan upon which creation was built.

The first figure that demonstrates force manifesting as form is the triangle, or as a three dimensional solid, the tetrahedron, a three sided pyramid of equilateral triangles. But probably the most important basic, building block of sacred geometry is the vesica piscis, a pair of interpenetrating circles of equal size, both of which have their centres on the other circle's circumference. In THE CANON, published by RILKO, William Stirling states:

> **"This mysterious figure, the Vesica Piscis, possessed an unbounded influence on the details of sacred architecture and it constituted the great and enduring secret of our ancient brethren."**

Illustration 68: The vesica piscis, the most important basic construction of sacred geometry.

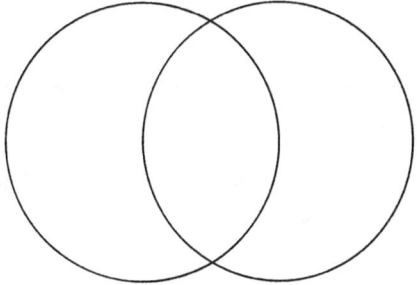

The reason for the importance of this deceptively simple construction may be that it holds a key to two of the most important figures in sacred geometry, the hexagram and the pentagram. The twin circles of the vesica intersect at four points, two at their centres, two on their circumferences. It has long been known that the intersectional points on each circumference identify a sixty degree division of the circles which aids the construction of a hexagonal figure (see Illus. 70). Less well known is that the other two points, the centres, maybe used to construct pentagonal figures.

To do this, the vesica must be contained within an outer circle, a step which may be taken symbolically to indicate that the vesica's progression from unity to duality is an internal division within the 'one' rather than a separate external development. Once the outer circle has a diameter drawn along the vesica's axis, it may be used to create a pair of interpenetrating pentagrams (Illus. 71). The initial angle created by this method is not precisely 36 degrees so the pentagram is not absolutely perfect. However it is close enough to allow an acceptable construction without recourse to protractors, which our ancestors did not possess.

You'll no doubt recognise this design as part of the Earthstars

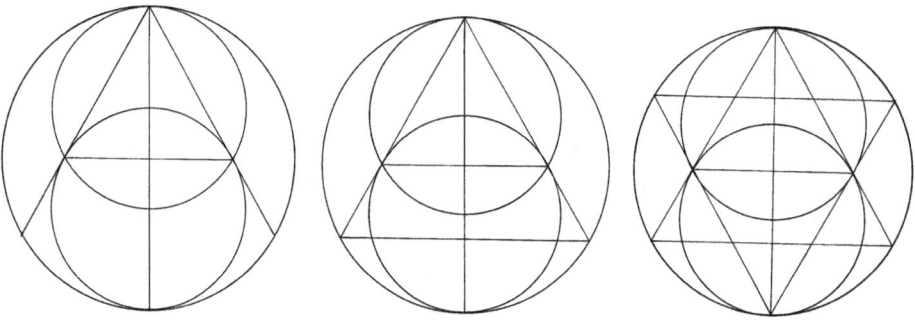

Illustration 69: The vesica's intersecting lines demonstrate a hexagonal relationship.

Illustration 70: Less well known is the vesica piscis' embodiment of pentagonal symmetry.

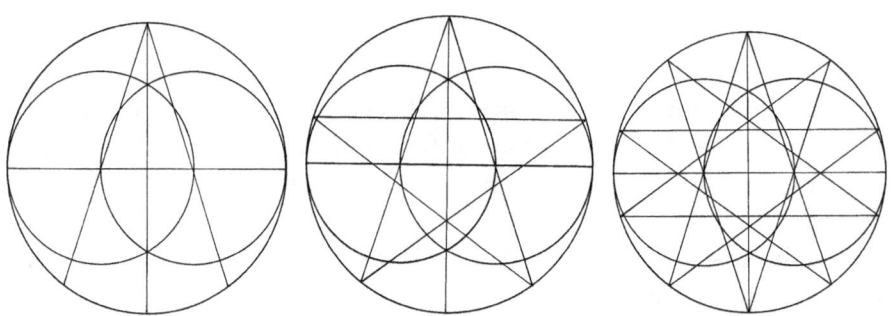

patterns upon the London landscape. If so, you'll appreciate how anyone with a fundamental knowledge of sacred geometry will have immediately recognised the basis of the Earthstars design.

Not only is it a well-known piece of sacred geometry, it is one of the most significant constructions in the whole subject. Some would say, the most significant construction.

THE SQUARED CIRCLE

At the time I stumbled across this discovery, my knowledge of sacred geometry was minimal. I simply recognised a similarity between the patterns on the London landscape and those I had seen in one of John Michell's books, CITY OF REVELATION.

The design turned out to be the culmination of a process known as squaring the circle and it encapsulates the very essence of sacred geometry since it is taken to be a graphic illustration of how divine force manifests in the material world.

The circle is traditionally a symbol of the eternal spiritual realms, the infinite, the universal and intangible. It has no beginning and no end and despite being a single, self-contained unit, encompasses everything. Any calculation of its dimensions reflects its infinite essence, being based upon Pi, one of mathematics' ultimately incalculable figures.

The square, on the other hand, represents the physical world of harsh reality, the four corners of the Earth, the four points of the compass, the four seasons, the four alchemical elements of Earth, Air, Fire and Water, which in turn may be thought of as the four physical states of matter; solid, liquid, gaseous and energy. In contrast to the circle, its dimensions are all easily calculable, finite measures.

Graphically, squaring the circle entails the reconciliation of these two apparent opposites by creating a square equal to a given circle, either in area or circumference.

Symbolically, it represents a union between heaven and Earth. In consequence, it was sometimes used as a temple ground plan, to create a sacred structure within a plot of hallowed ground by using the divine plan to link heaven to Earth. The Pantheon in Rome is just one example.

That London's landscape geometry is based upon identical foundations to the squared circle is undeniable. The proportions of their concentric circles are identical and in sacred geometry, it is proportion rather than scale that is the important factor. London's two circles are

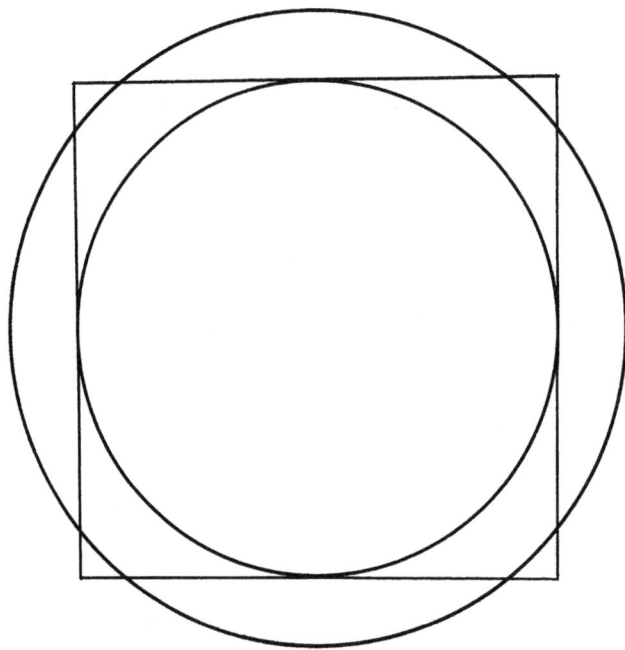

Illustration 71: The basic outline of the squared circle, its two concentric circles proportionately identical to those in London's Earthstars geometry.

16.2 and 20.625 miles respectively, so matching the 1;1.27272 ratio of the squared circle diagram remarkably closely.

In addition, two methods of constructing a squared circle evolve from geometry almost identical to London's Earthstar patterns.

Any two adjacent points of a pentagram can be used to create the base line of a square on the pentagram's circle. Naturally, additional pentagonal figures such as London's patterns could provide the basis of the square's other lines. (Illus 72-73). According to Robert Lawlor in his book, SACRED GEOMETRY, PHILOSOPHY AND PRACTICE, this construction originates from the middle ages and is not mathematically exact although he states that the resultant squared circle is roughly equal in area to the initial circle, within acceptable parameters.

London's concentric circles and pentagram could form the basis of such a figure but there's another construction which is a little more accurate and far more applicable to London. It's a squared circle obtained from a double pentagram or decagon on the inner circle, exactly like those to be found amongst the London Earthstar patterns.

Illustration 72: The squared circle constructed from the base of a pentagon.

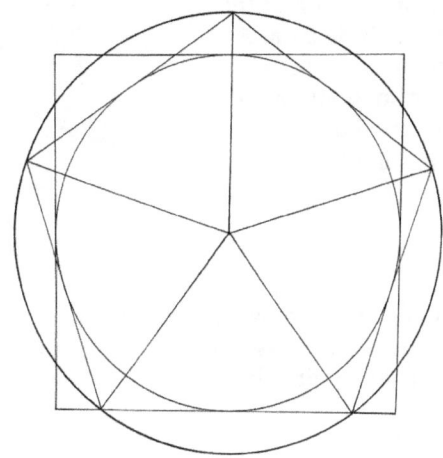

Illustration 73: The construction of this squared circle utilises a twenty-point star like the one I found on the London landscape.

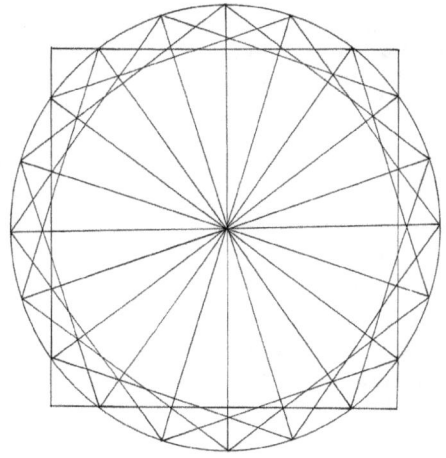

To construct it from a ten-point star, centre a compass on any point and set a radius to three points further round the circumference in either direction. Then draw an arc across the circle through these points.

Repeat the process from the point diametrically opposite your first. Where the two arcs intersect will define the diameter of the outer circle. It is then a simple matter to complete the figure by constructing a square around the original (inner) circle that will be equal in perimeter to it.

When this square is superimposed over London's other patterns it turns

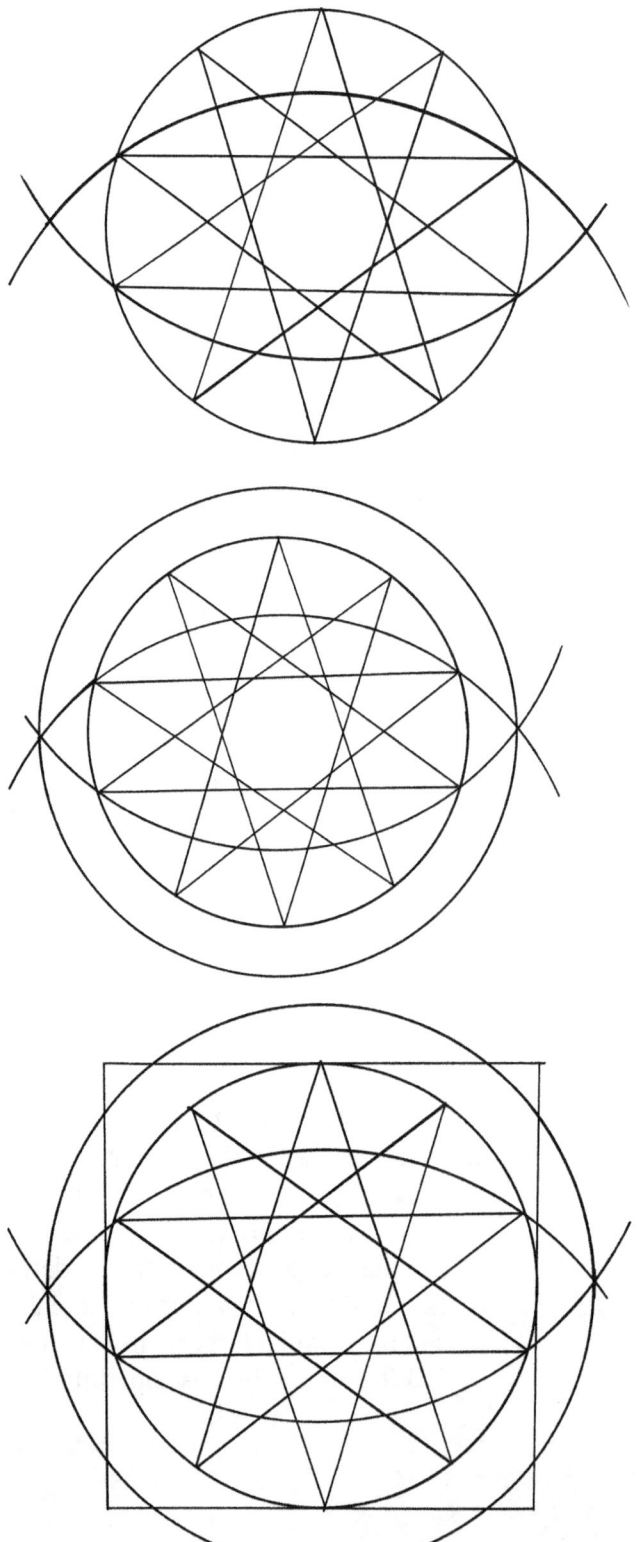

Illustrations 74, 75 & 76: The construction of the squared circle from a double pentagram or decagram identical to those found on the London landscape.

First centre a compass on point A and draw an arc through points B and C.

Then centre a compass on point D and draw an arc through points E and F.

From the centre of the circle, take a radius to either point where the arcs intersect. This produces the correct proportional relationship between the two circles.

It is then relatively simple to construct a square upon the original circle to complete the squared circle.

up some very significant sites. The N.E. corner is Grimstone's Oak, a well-known landmark at the junction of four paths in Epping Forest. The N.W. Corner is a hilltop point near Grim's Dyke, near Harrow Weald Common Wood, marked by an O.S. triangulation point. The S.W. corner turns out to be a spot marked on the map as Seething Wells, near Esher, although if the wells there still seethe they are the property of Thames Water and not accessible to the public. The S.E. corner is in the grounds of Chislehurst Golf Club and while I could only find grass and large oaks in the vicinity, I have since been informed that a well or spring lies near to the spot. In days gone by, hilltops, wells, springs and venerated trees all served to mark sacred sites or shrines, so these four places may embody a spiritual presence that has so far gone unrecognised by today's materialistic society.

Since the squared circle geometry and London's Earthstars share a common basis and many identical features, their symbolism may be interpreted in the same way, allowing us an insight into the significance of this discovery.

In general terms, any construction of sacred geometry can be interpreted as an expression of the formative forces of creation. London's geometry is no exception.

The square is of course, the square of the Earth.

Circles are representative of spirit or of perfected nature, so the circle contained within the square may be defined as the Circle of the Earth Spirit. The outer circle is the Circle of the Heavens, or in this case we could call it the Circle of the Universal Spirit, or Holy Spirit.

Of course, the star patterns within the circles and squares each have their own symbolic interpretation.

The pentagram for instance, shares the circle's associations with infinity because, like the circle, it can be drawn with one continuous unbroken line. It may also be said to represent spiritual dominion, or the influence of spirit over the world of the four elements.

In THE BOOK OF DRUIDRY by Ross Nichols, four of the five points are associated with the four elements, Earth, Air, Fire, Water. The fifth and uppermost point represents spirit. These attributions transpose remarkably well onto their respective sites of the Earthstars pentagram. The Fire point is Bellingham Green which may derive its name from associations with Bel, Belinus or Baal, all names associated with sun gods. The Earth point is an earthworks, Caesar's Camp on Wimbledon Common. Air is the highest point, Horsenden Hill. The water point is St. Gabriel's church near Wanstead Flats and in some esoteric traditions, the Archangel

Illustration 77:
Pentagram with elemental points, from Ross Nichols'
THE BOOK OF DRUIDRY.

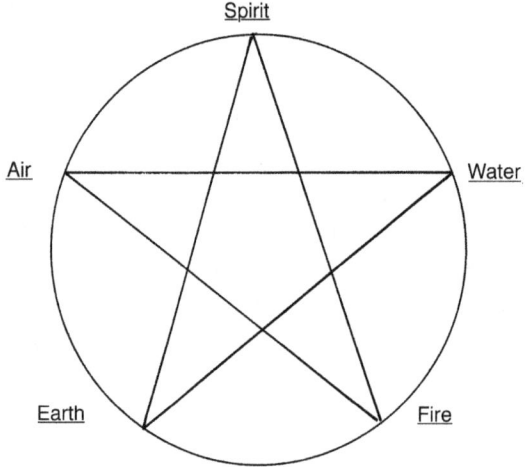

Gabriel is associated with the element of water. More remarkably, at the fifth point, which corresponds to spirit, we find a huge symbol for spirit emblazoned across the landscape of north London; the Barnet Triangle.

This is, of course, an equilateral triangle traditionally said to symbolise the Holy Trinity, or Holy Spirit. If we continue this line of thought, St. Mary's in East Barnet becomes the place where the Holy Spirit enters the circle, which in a way, confirms the message of the 'spiritual power station' dream.

The squared circle's association with cosmic relationships is not restricted to mere symbolism. Sacred geometry is founded upon a detailed analysis and identification of the geometric forms that underlie all creation. The way its principal patterns develop and fit together literally reflect the order and structure of the universe. Nowhere is this more apparent than in the construction of the squared circle whose proportions illustrate perfectly a very specific relationship between the Earth and the heavens above it.

When the Circle of the Earth Spirit is given the exact proportions of the Earth, a mean radius of 3,960 miles, the distance to the outer circle of the heavens is 1,080, the mean radius of the Moon.

Solar dimensions are present in the diagram too, although less overtly. The perimeter of a square around the lunar orb would measure 8,640 miles, a figure directly associated with the sun whose mean diameter is 864,000 miles.

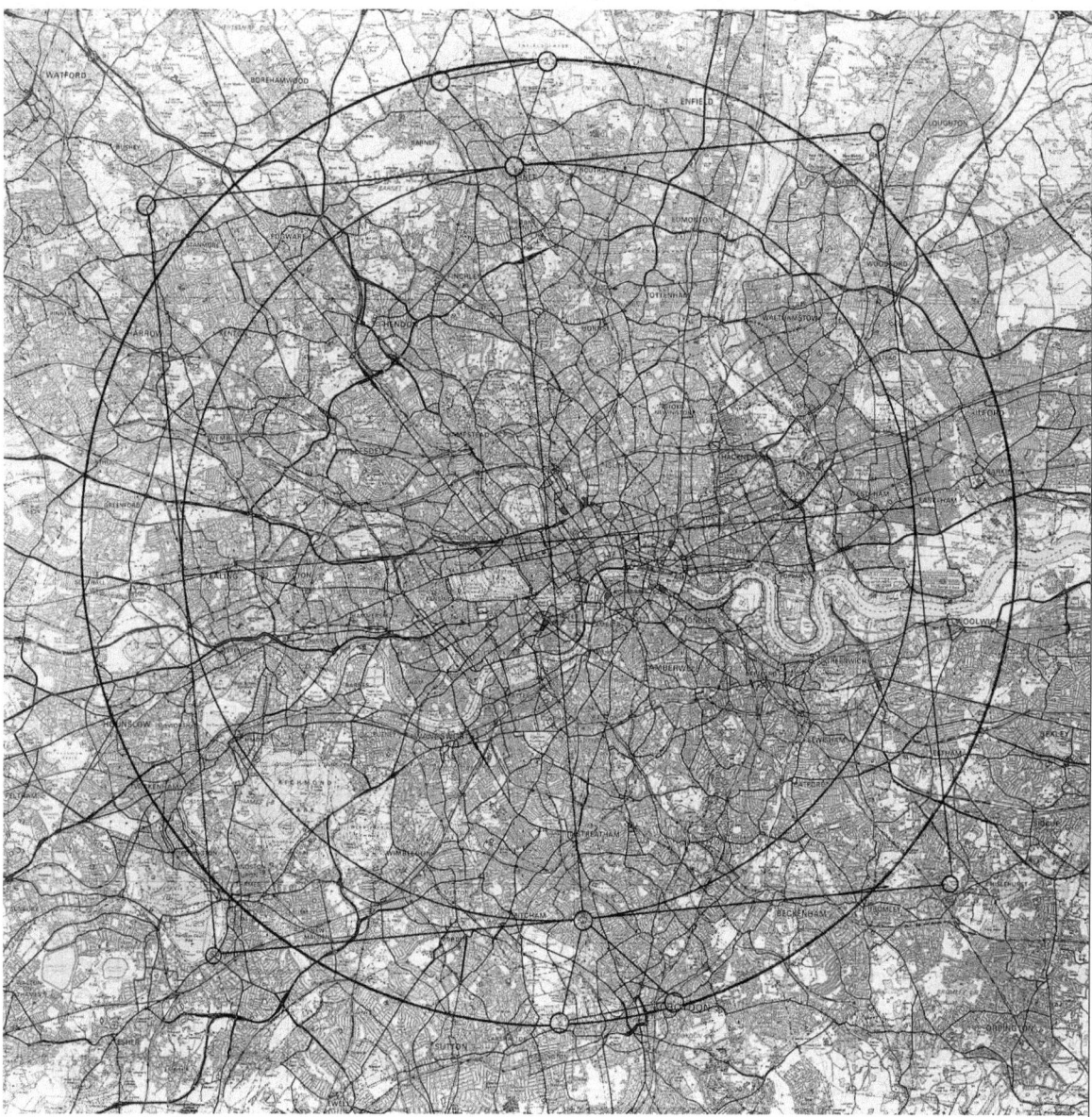

Illustration 78: The basic squared circle diagram superimposed upon the London area. O.S. map © Crown Copyright. All rights reserved. License No. 100029558.

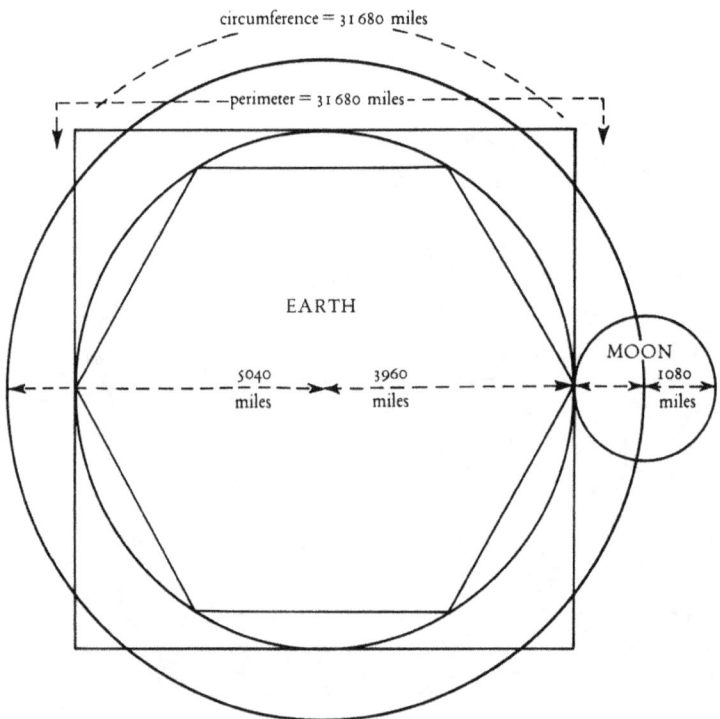

Illustration 79: The squared circle embodies the proportional relationship of the Earth and Moon. Anything built upon this pattern will resonate holistically with the planet (reproduced from John Michell's CITY OF REVELATION with permission).

HARMONIC NUMBERS

All of these numbers recur with alarming regularity in the actual proportions of this figure upon the London landscape.

The Earthstars' Circle of the Earth is 16.2 miles in diameter. That is 28,512 yards, or 3,960 x 7.2 (Earth's radius number x pentagonal angle number); Or 216 (lunar diameter number) x 132, or 33 x 864 (Sun's diameter number).

In feet, the figure becomes 85,536, which is 108 x 792 (lunar radial number x Earth diameter number), or 99 x 864 (sun's diameter number).

In inches, we arrive at 1,026,432, or 396 x 864 (Earth radius number x Sun's diameter number), or 3168 x 1080 (lunar radius number).

In furlongs, this gives 129.6, which equates with 10.8 (lunar number) x 12, or 864 (solar number) x 1.5. In poles (16.5ft), it's 5184, or 48 x 108 (lunar number), or 6 x 864 (solar number).

Similar numbers recur in association with the circle's circumference which works out initially to an unspectacular-looking 50.914285 miles.

In inches, that becomes 3,225,929, a seemingly random and cumbersome number until you divide it by the equally cumbersome 24,883.2, the Earth's mean circumference in miles. The result is another recurrent number, 129.6, which just happens to be 10.8 (again) x 12, or 3 x 43.2 (solar radius number).

These astonishing correspondences are not restricted to the proportions of just the circle; They recur in a great many more of the figures, including the Barnet and Croydon Triangles, the decagonal star, the eight-point star and the square of the Earth.

The Barnet Triangle's 7.2 miles perimeter converts to 12,672 yards, which is 1.6 x 7920 (the Earth's mean diameter number); In feet, it is 38,016 which is 44 x 864 (the solar diameter number) or 48 x 792 (Earth diameter number).

The Croydon Triangle's side from Pollard's Hill to St Mary's Beddington is 2.25 miles; which is 3,960 yards, the same figure as the Earth's mean radius in miles. It's also 720 poles (the pentagonal angle number again). As 142,560 inches, it is 432 (solar radius number) x 330.

The Decagon's sides are exactly five miles. That's 8,800 yards or 26,400ft.This pattern can also generate similar correspondences with the very same numbers; 8,800 is 1.11111 x 7,920; 26,400 is 7,920 x 3.333r or 6.666r x 3,960.

The decagonal star's chords produce yet more repetitions of these familiar figures. They are 12.96 miles precisely. In yards, that's the insignificant looking number 22,809.6. However, a few calculations reveal it is actually extremely significant in that it is also 288 x 79.2, or 7.2 x 3168, or 864 x 264, or 864 x 4 x 6.6. In feet, it's 68,428.8, or 864 x 792, or 21.6 x 3168.

Not to be excluded from all this, the eight-point star's chords are exactly 15 miles, or 79,200ft.

The Earthstars' Square of the Earth is equally productive where symbolic numbers are concerned. Its perimeter is 64.8 miles which is 518.4 furlongs or 21.60 leagues or 114,048 yards. 114,084 is 144 x 792 and 5184 is 72, the pentagonal number, squared.

If nothing else, this confirms that the Earthstars' square of the Earth has a firm foundation in the reality of numerical relationships.

I could go on, since there are many more examples of these numbers to draw upon, but it would become tedious.

The explanation for these correspondences is in the nature of the geometry itself which embodies universal structure. If the sacred geometry is accurate, significant numerical proportions will always be built into it, provided the measures involved, in this case our ancient British metrology, are derived from a tradition of which sacred geometry or an understanding of the foundational principles was a part. The holistic relationship to the whole is then automatically built in.

In the traditional British measures, this is obvious. There are 7,920 inches in a furlong. 792 inches in a chain. A mile is 792ft x 6.666r. A league is 792yds x 6.6666r. All extremely appropriate figures for measures of the Earth.

Significant numbers also occur in the phenomenon known as the precession of the equinoxes - the apparent rotation of the zodiacal constellations caused by the Earth's axial tilt. Also known as the Great Year, it consist of twelve great months, each lasting for 2,160 years (the lunar diameter number). Each day of the Great Year is 72 of our years (the pentagonal angle number). Four months would be 8,640 years (the solar diameter number).

These same numbers are equally evident in the units we use to measure time, which, of course, here on Earth derives its periods directly from the relative cycles of the Earth, Sun and Moon.

There are 86,400 seconds in a day, 864,000 miles in the mean diameter of the Sun: 43,200 minutes in a month, 432,000 miles in the radius of the Sun; 2,160 hours in a quarter year, 2,160 miles in the diameter of the Moon: In this context our twelve "moonths" are something of a compromise since there are actually thirteen lunar months annually.

The Sumerians are generally credited with creating our units of time, supposedly long before anyone was capable of calculating the Sun's mean diameter or anything remotely similar.

Like all traditional units of metrology such as the inch, the foot, the furlong, the rod, the mile and so on, seconds, minutes and hours are units of cosmic proportion utilised to ensure that the works of man were in time and scale with the workings of the universe and its creator.

The Earthstars' geometry is clearly an intrinsic part of the Earth's structure and since they are both built upon the golden mean proportion, each part automatically shares a harmonious relationship with the whole. Suffice to say, the Earthstars mandala resonates in harmony, not just with the planet, but with the universe at large.

This is not simple geometry, but the divine designs of sacred geometry The forces it represents may be understood as a manifestation of the formative forces of creation, the life force and inner, spiritual energies of the planet itself.

They would have been present ever since the Earth cooled and solidified from a blob of solar plasma, just as a droplet of airborne moisture displays the geometry of its inner structure as it cools and solidifies into a snowflake, or as the molten lava which formed the devil's causeway in Ireland solidified into hexagonal blocks as it was cooled by sudden contact with the sea.

From this point of view, designs like the Earthstars covering London would have to be regarded as a perfectly natural phenomenon rather than the result of any monumental human effort.

Sadly for the Von Daniken fans out there, that might also rule out any theories involving alien visitors, landscape gardeners from outer space or little green Capability Browns.

However, the truth might actually be even more far-fetched. Since they share the same symbolism and structure, the inference is that the squared circle and London's Earthstar patterns share similar functions.

There is of course, one big difference.

Sacred geometry hidden within the landscape and traceable through the location of certain sacred sites cannot be merely symbolic.

It may also be functional.

It may actually be some kind of circuit diagram of power in the land. If so, it is directly associated with the life force of the planet.

The ultimate power source.

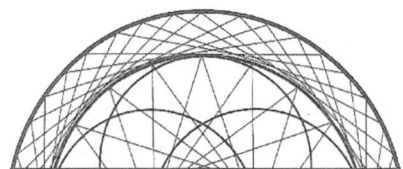

Chapter Eight
The Full Circles

It seems remarkable to me that every single one of the patterns so far have been generated by two relatively simple triangles, the Barnet Triangle to the north of London and the Croydon Triangle at the south.

With the development that the east and west points were also marked by significant features or at least had been in the past, it seemed sensible to see if they, in turn, revealed any interesting triangulations of sites. My initial attempts at this took the form of map dowsing with a pendulum. Three sites in East London gave a positive reading, one in Barking, one in Ilford, one in East Ham. A recce to the area confirmed that all three sites were of great antiquity and each had a very different feel to the atmosphere.

THE EAST HAM TRIANGLE

The Ilford site turned out to be the old St. Mary's Church at Seven Kings. It was an extremely ancient place of worship and although it is a potent spot with some very interesting connections to other sites, on that day, it simply felt like the wrong place to be, so I headed for Barking. Here, I found the church I had dowsed was dedicated to St. Margaret. What the map hadn't shown was that it stood next to the ruins of Barking Abbey and up until that point, I didn't even know Barking had an Abbey.

As I strolled into its grounds, the sense of energy was so powerful in places that my hands throbbed with it, rather than responding with the usual tingling sensations. There was also a strong sense of the White Lady's presence within the spirit of the place, particularly in the area of the Abbey ruins, which lie in a hollow next to the church. All that's visible now is the foundations of the main walls, a ground-plan drawn in stone upon the green lawns. The only other surviving piece of the original abbey is the gatehouse at the eastern entrance. A commemorative plaque on its wall recounts the abbey's history.

Barking Abbey was founded in 666 AD by Erkenwald who later

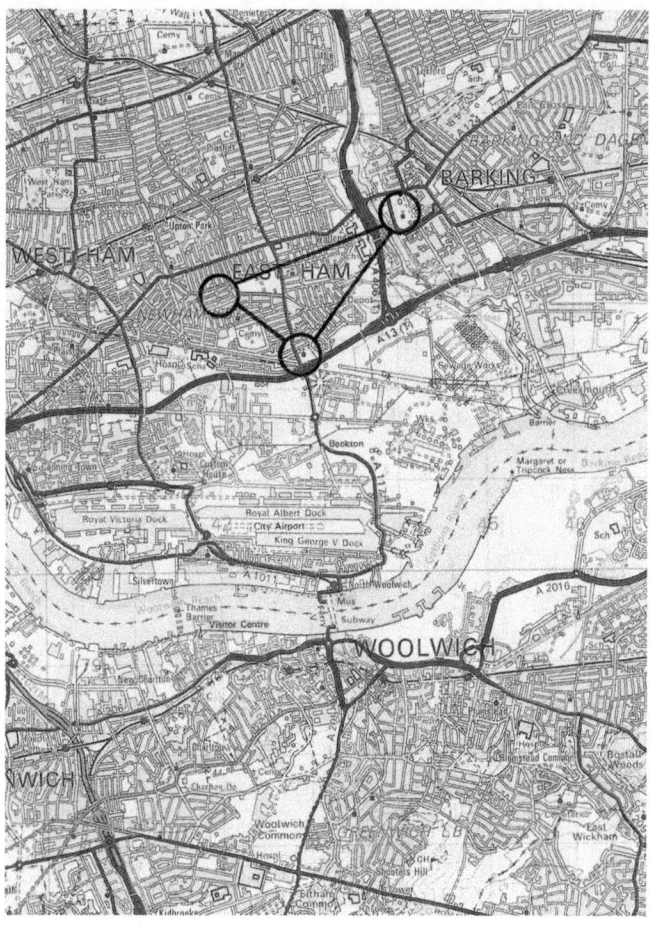

Illustration 80: The East Ham Triangle formed by the locations of Barking Abbey, East Ham's Parish Church of St. Mary the Virgin and the eastern point of the Earthstar's circle, the lost Miller's Well. O.S. map © Crown Copyright. All rights reserved. License No. 100029558.

became Bishop of London. His sister Ethelburga was the first of a long line of Abbesses. It had been sacked and pillaged by Danes in 870, then rebuilt, only to be destroyed in 1541 following Henry VIII's dissolution of the monasteries, at which time it was the richest abbey of its kind in England, surpassing even Westminster in wealth.

In its day, it was obviously an important spiritual centre, possibly of greater significance than Westminster Abbey which, coincidentally, also has a St. Margaret's church right next door and is linked to Barking Abbey by an alignment that passes through the Earthstars' pentagonal junction point on Tower Hill, through the Tower itself and through Southwark Cathedral (further information on this alignment is to be found in a later chapter).

A visit to the East Ham site was equally rewarding. Here, I found that the old parish church was another unique sacred site of great antiquity: the only Norman church in London to have survived virtually intact and once again, it was dedicated to St. Mary the Virgin.

The churchyard looked a little overgrown but only because nature had reclaimed it. It was now officially a nature reserve and life appeared to be thriving amongst the gravestones. The atmosphere here was exceptional and the presence of the White Goddess was almost tangible amidst the lush undergrowth. I asked the nature reserve wardens if I could see inside the church, which retains many of its original features and was allowed access. It was not in regular use as a place of worship, but still used for services on special occasions.

Inside, the atmosphere sang, literally. The energy of the place was so intense the whole building seemed to hum. I sat silently absorbing it for as long as I could, then left somewhat dazed, but invigorated by the experience and convinced I had found another spiritual power point in the Earthstars' network.

The East Ham Triangle was defined by Barking Abbey, the Parish Church of St. Mary The Virgin East Ham and the Earthstars' circle's eastern point, the Miller's Well, the now-forgotten therapeutic spring or well beneath Central Park Road. Given this set of starting points, no amount of over-drawing on maps could make a triangle resembling the Barnet or Croydon ones. It was a triangle all right, but not an equilateral one. It struck me as a being a distinctly odd little shape.

Astonishingly, the oldest sacred sites on the western side of the Earthstars circle produced a mirror image of it.

THE HANWELL TRIANGLE

Initially, I had no inkling that these were the oldest churches in the area. I was simply dowsing places on the map, trying to identify a possible triangle focused on the western point of the Earthstars circle. The quest led me to the parish churches of Hanwell and Norwood Green. Again, both turned out to be dedicated to St. Mary and both were of considerable antiquity.

St. Mary's at Norwood Green was the Mother church of Southall, built around 1200 AD probably upon the site of an earlier place of worship. It's a picturesque spot, a small oasis of rural calm beside a busy suburban road.

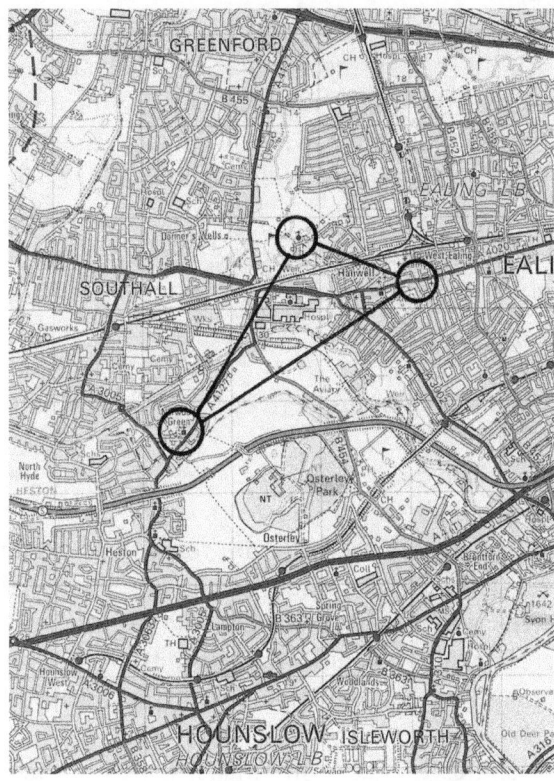

Illustration 81: The Hanwell Triangle, a mirror image of the East Ham Triangle and formed by the location of the oldest sacred sites in this area; St. Mary's Hanwell, St. Mary's at Norwood Green and the western point of the Earthstar's circle, near the former site of a boundary stone. O.S. map © Crown Copyright. All rights reserved. License No. 100029558.

St. Mary's at Hanwell has an even lengthier history, which was tracked down for me by John and Joan Taplin who live in the area. The village records suggest a place of worship may have served the village since Saxon times and that it may possibly have sprung up around a venerated well recorded as early as 588 AD (BYGONE HANWELL, Chapter 3) as "Hanan-wel" or in 998 as "Hana-wella" - the well of Hana. Hana may well be a local name for the Mother Goddess who is known elsewhere as Ana, Anu, Dana or Danu.

The most likely location for Hana's well is a spring, which the Taplins informed me, could still be seen in the grounds of a private house not far from the church. As Margaret Woods observed in relation to East Ham's healing waters, a venerated spring or well in ancient times would have given rise to a shrine in the immediate vicinity, probably, as in this case, dedicated to a local goddess.

The goddess of these waters may be faintly remembered today in an inscription carved into an arch over the small stream flowing from the spring;

"Dulcissimo Sonorum pari amoris ergo et hospitii hos fontes."

The waters are therefore dedicated to the sweet sisters of the spring or well, for their love and hospitality.

Who the "sisters" referred to actually are remains a mystery. It may relate to some ancient sisterhood who tended the spring, to the naiads of the waters or to the ancient goddess of the well who is still very much in evidence on the psychic levels. A clairvoyant vision of the white goddess drifting ethereally from the well towards the nearby pond linked this site to my other places of vision.

With two new triangles to add to my maps, it was not long before I began following the same procedures I'd used on the Barnet and Croydon triangles, tracing their extended sides to see what, if any, geometric shapes they might generate.

Illustrations 82 & 83: Extended lines from the East Ham and Hanwell triangles both produce pentagonal figures in the Earthstars' circle.

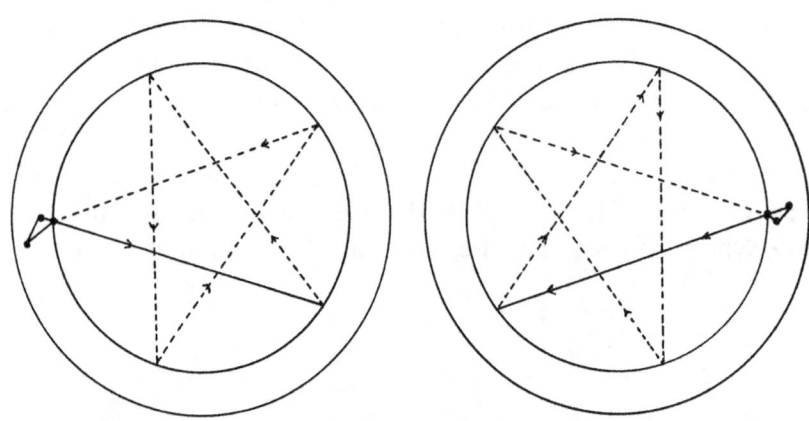

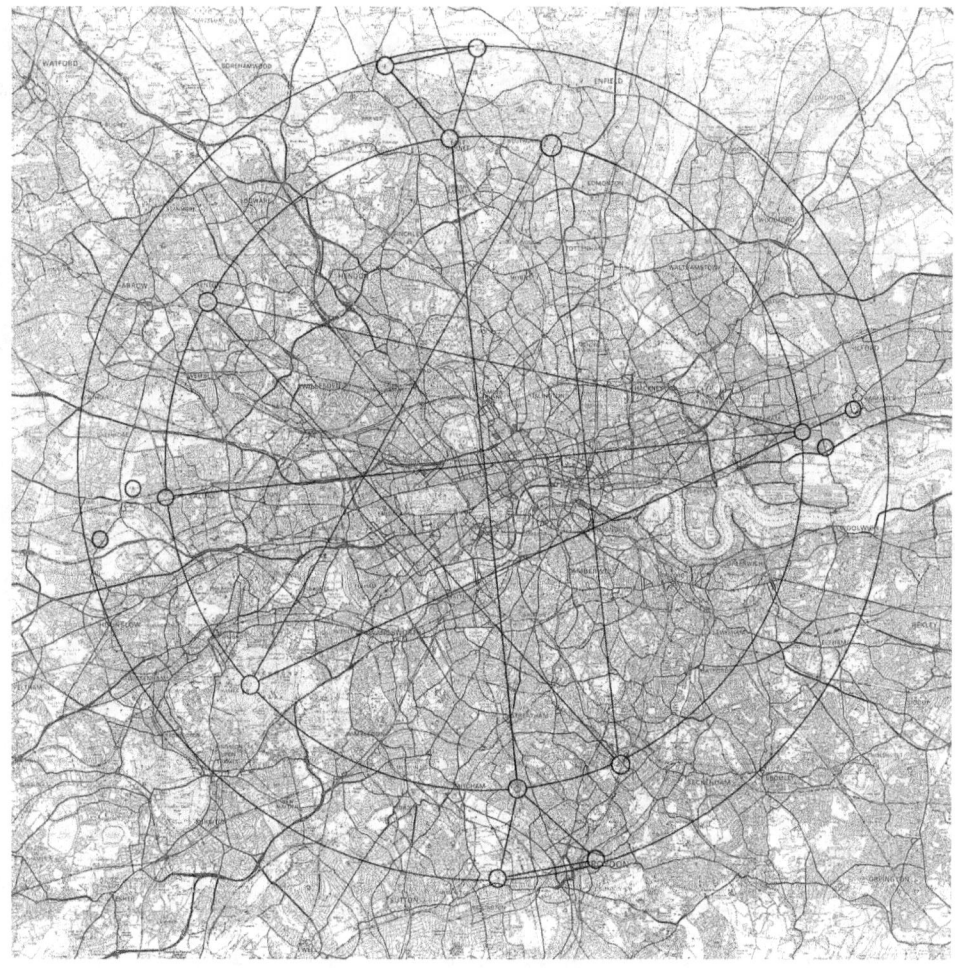

Illustration 84: Map of the pentagram generated by the East Ham Triangle. Clockwise from the eastern point of the circle, the mark points are; 1; The lost Miller's Well. 2; A point in a field near Annerley Town Hall. 3; Point in a wood in Richmond Park. 4; Hilltop in Kenton near Woodhill Crescent. 5; Point in Station Rd Winchmore Hill. O.S. map © Crown Copyright. All rights reserved. License No. 100029558.

I could hardly believe the shapes that evolved. Not only did each triangle create a distinct and perfect pattern, they both created the same kind of pattern; Five-pointed stars. The Eastern pentagram derives from an extension of the line from Barking Abbey through the eastern point of the circle. The Western pentagram from a similar extension of the line from St. Mary's Hanwell through the Western point. Of course, together they make another ten-point star, this time aligned to the East-West axis rather than the North-South one.

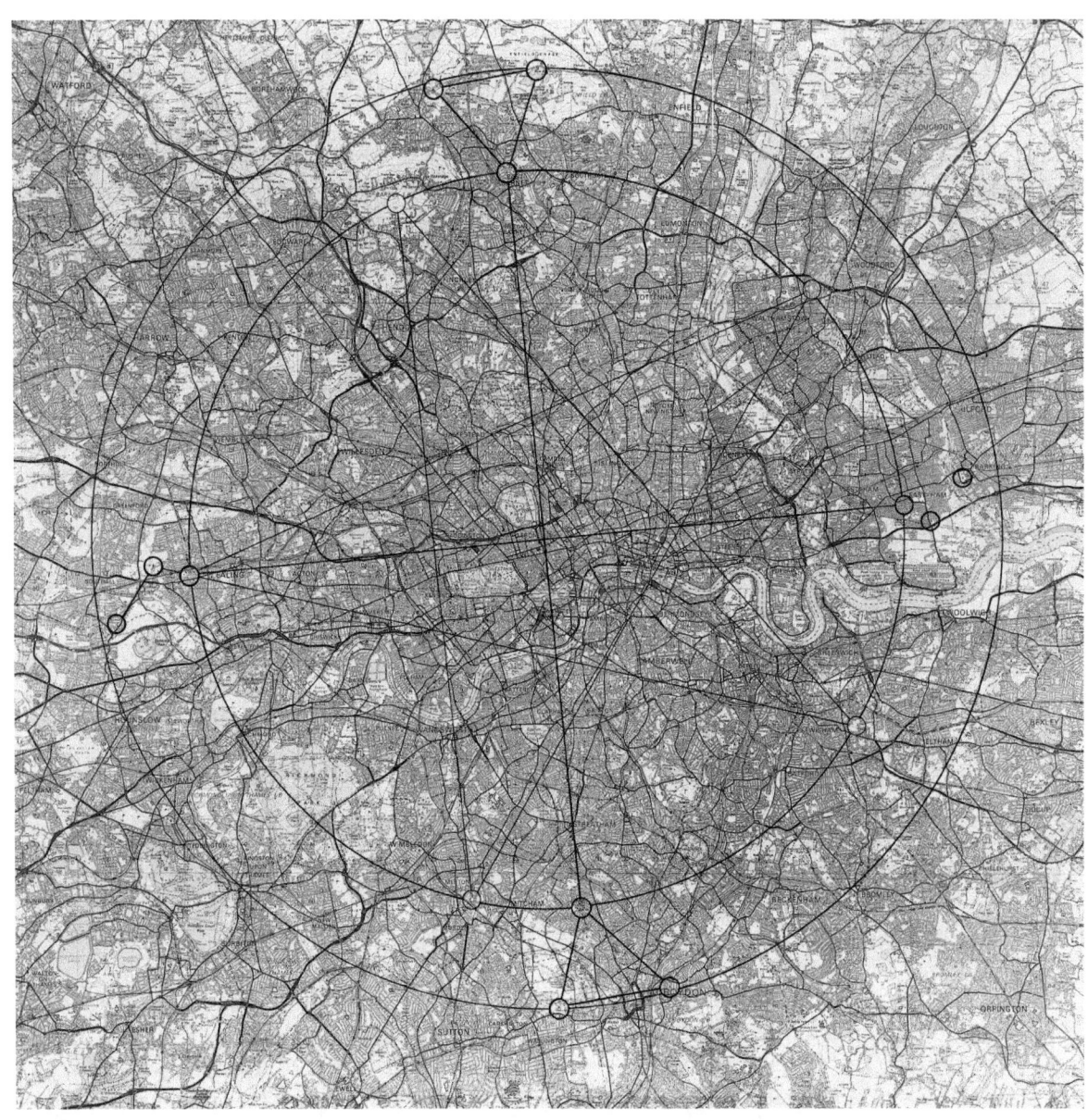

Illustration 85: Map of the pentagram generated by the Hanwell Triangle. Mark points clockwise from the western point of the circle are; 1: The site of a boundary stone. 2: Point in fields between Burtonhole Lane and Totteridge 3: Point in Hale End Rd, E17. 4: Point in Weigall Rd Sports Ground, Eltham. 5: Point in Rock Terrace Recreation Ground, SW19. O.S. map © Crown Copyright. All rights reserved. License No. 100029558.

There are several notable mark points on both stars; a number of churches, of course; several hilltop sites like the area near Crystal Palace's mast where huge stone sphinxes gaze out over the landscape; the hilltop in Finsbury Park again; the hills overlooking Highgate Ponds on Hampstead Heath; a junction of lines in Highgate Woods, close to the cafe.

A feature of the previous patterns crops up again, too, the parallel roads. This is far more noticeable on the western pentagram where one of its lines, running parallel to the A5 Edgware Road, also directly traces the course of the Holloway Rd almost all the way from Holloway to Shoreditch. It made me wonder if Holloway Road might not derive from an ancient Holy Way or pilgrim's route but that raises the question of where it leads. This straight section points directly towards Shoreditch Parish Church which has the remains of a very ancient mound nearby at the centre of Arnold Circus. In the opposite direction, of course, the road takes you to London's highest hill at Highgate, equally valid as an ancient holy place.

Another line of the same pattern parallels the A3 then the A24 for several miles, all the way from the Elephant and Castle to Merton.

Like the pentagrams which evolved from the Barnet and Croydon triangles, these are a perfectly interlocking matched pair which join together to create a ten-pointed star on the London landscape.

I had thought it beyond the realms of coincidence that the Croydon and Barnet Triangles both produced pentagonal figures which proved to form an interlocking pair. To find two more triangles that do exactly the same and are mirror images of each other is startling evidence for the validity of this discovery.

THE 20 POINT STAR

Naturally, the east and west five-point stars link directly with the previously discovered pentagonal patterns to create a far more impressive figure. The full pattern is a twenty-pointed star. It naturally has two ten-point stars within it, but also comprises four pentagrams, each orientated towards one of the four compass points, north, south, east and west.

In terms of the squared circle geometry, the circle they lie on is the Circle of the Earth Spirit and as a symbol of hidden powers within the Earth, four pentagrams, one directed towards each of the cardinal and elemental points make an extremely potent and appropriate image.

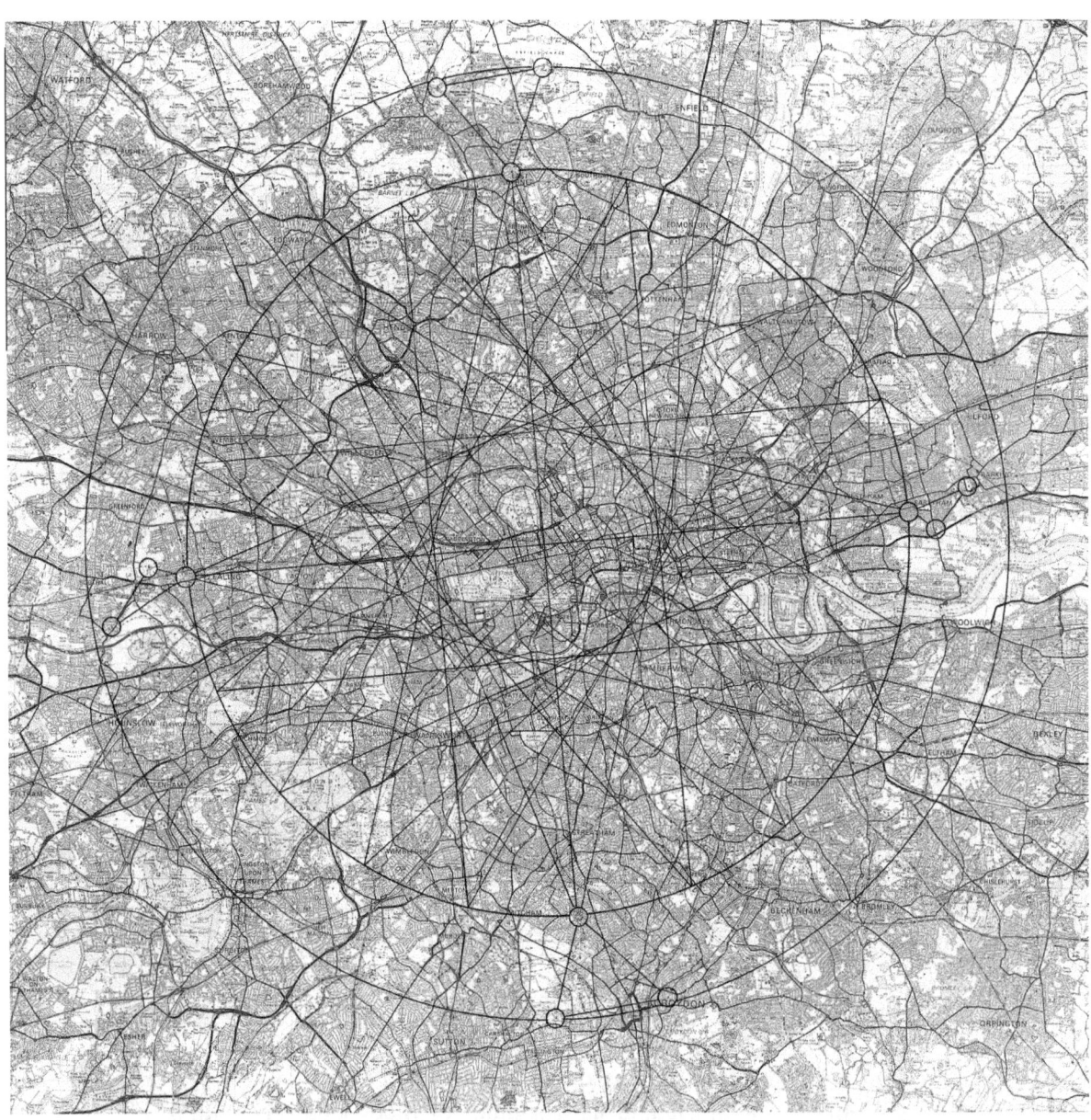

Illustration 86: The Circle of the Earth Spirit, containing a 20-point star composed of four pentagrams, each one directed towards one of the four compass points. O.S. map © Crown Copyright. All rights reserved. License No. 100029558.

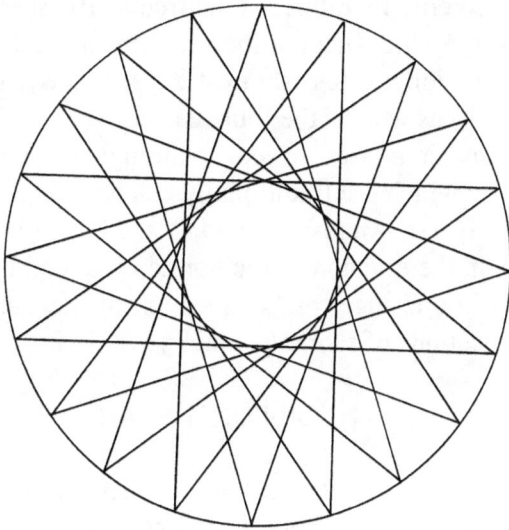

Illustration 87: The 20-point star composed of four pentagrams, aligned to the four compass points.

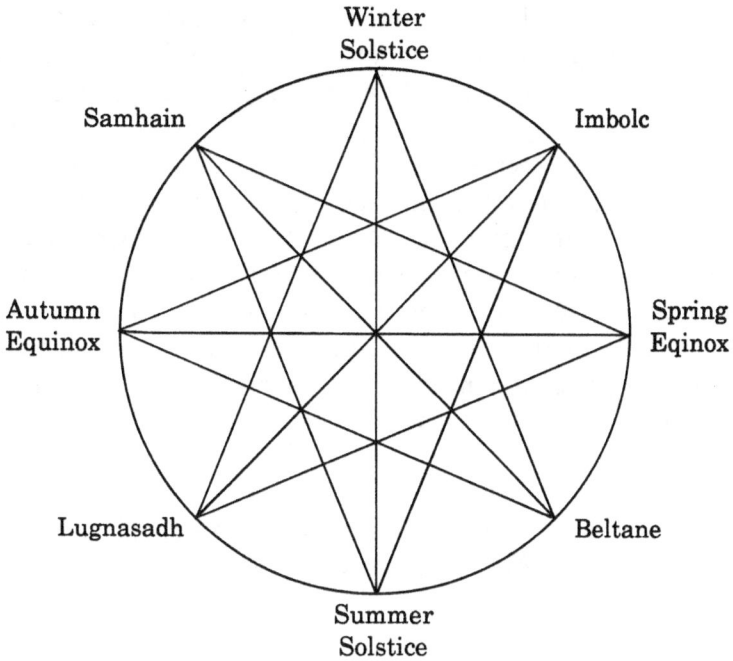

Illustration 88: The eight-point star as the wheel of the year, the traditional eight festivals which mark the turning points of the year and the cycle of the seasons.

The eight-pointed star which also falls on this circle is equally appropriate. In esoteric literature, it is frequently seen representing the cycles of the Earth. As the wheel of the year, it is naturally divided into four by the solstices (the longest and shortest days) and equinoxes (the days of equal light and dark) as well as the four seasons.

The eight divisions are arrived at through the addition of the four Celtic fire festivals which are traditionally celebrated on or near the quarter days, Feb. 1st, May 1st, August 1st and Oct. 31st. In addition, as we will see in Chapter Ten, the division of the horizon into eight segments played a fundamental part in our ancestors' conception of the sacred landscape and the practical applications of sacred geometry within it.

THE 30-POINT STAR

Extending the sides of the earlier triangles had provided remarkable results, so predictably, I set about a similar task with the East Ham and Hanwell Triangles to see if they identified any other new patterns, or confirmed any existing ones.

The Hanwell triangle provided just one more interesting figure; Extending the line from St. Mary's Norwood Green through the inner circle's western point creates a chord at an angle of 24 degrees to the radius. Extending it around the circle (at a reflected angle of 48 degrees) eventually gives rise to a complex thirty-pointed star on the outer circle.

Once again, additional conformation of the figure's validity is provided by an alignment from another triangle, in this case, an extension of the East Ham Triangle line from Barking Abbey to East Ham's St. Mary's.

The design this creates is a fifteen-pointed star on the same circle with a clear connection to the thirty-point star by virtue of the fact that its fifteen points are alternate points of the thirty. This phenomenon of one pattern being confirmed by a second related design has occurred before in the Earthstars discovery; in the original five and ten-point stars. Obviously, a fifteen-pointed star could also be drawn as three component five-pointed stars, so it is another design with a pentagonal basis. As a fifteen-point however, it was created from a single, continuous line reflected around the inside of the circle.

As with so many of the previous patterns, inherent within the construction of both the fifteen and thirty-pointed stars is clear evidence of their relationship to the existing figures.

London   City of Revelation

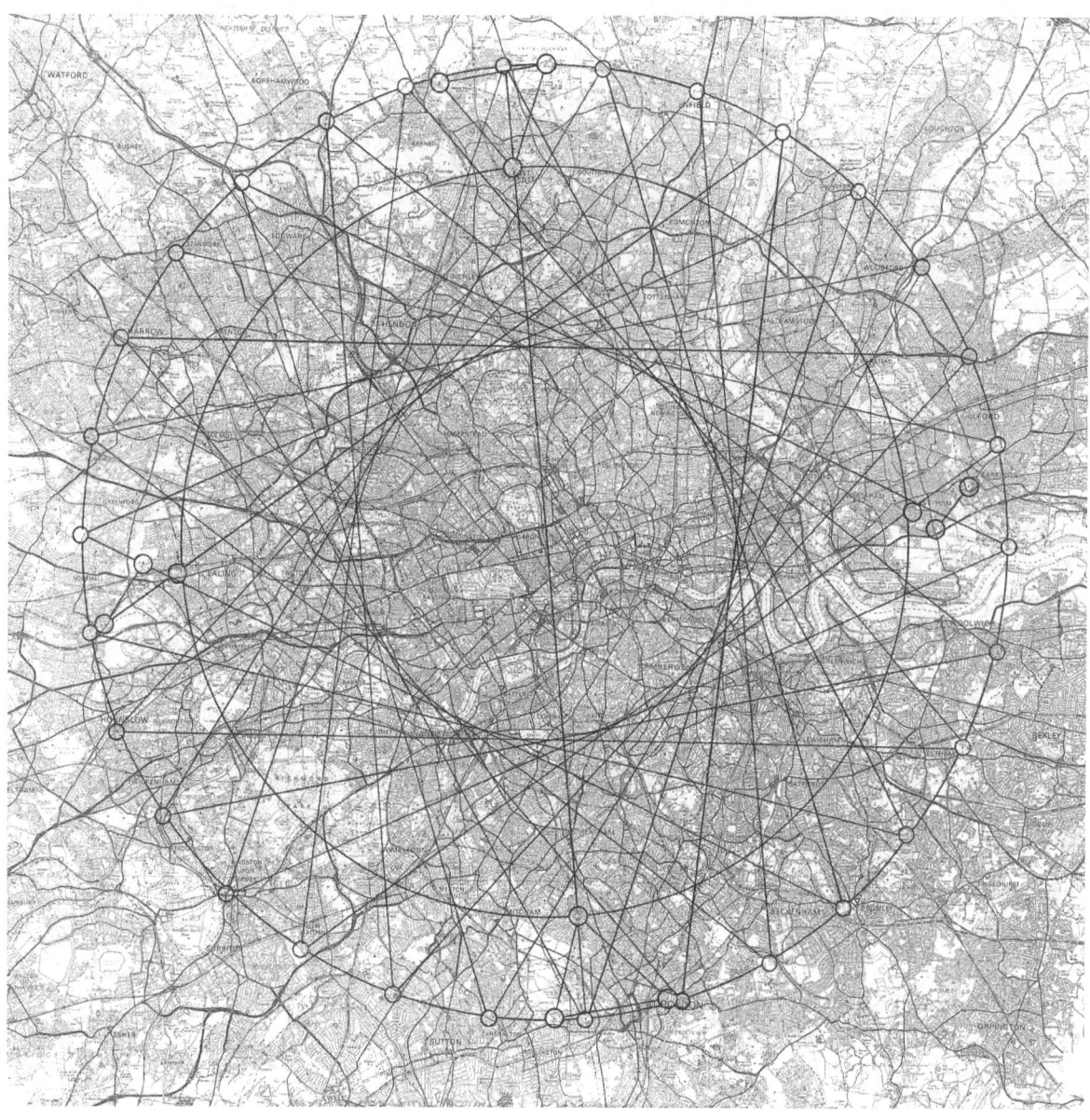

Illustration 89: Map of the complex 30-point star. O.S. map © Crown Copyright. All rights reserved. License No. 100029558.

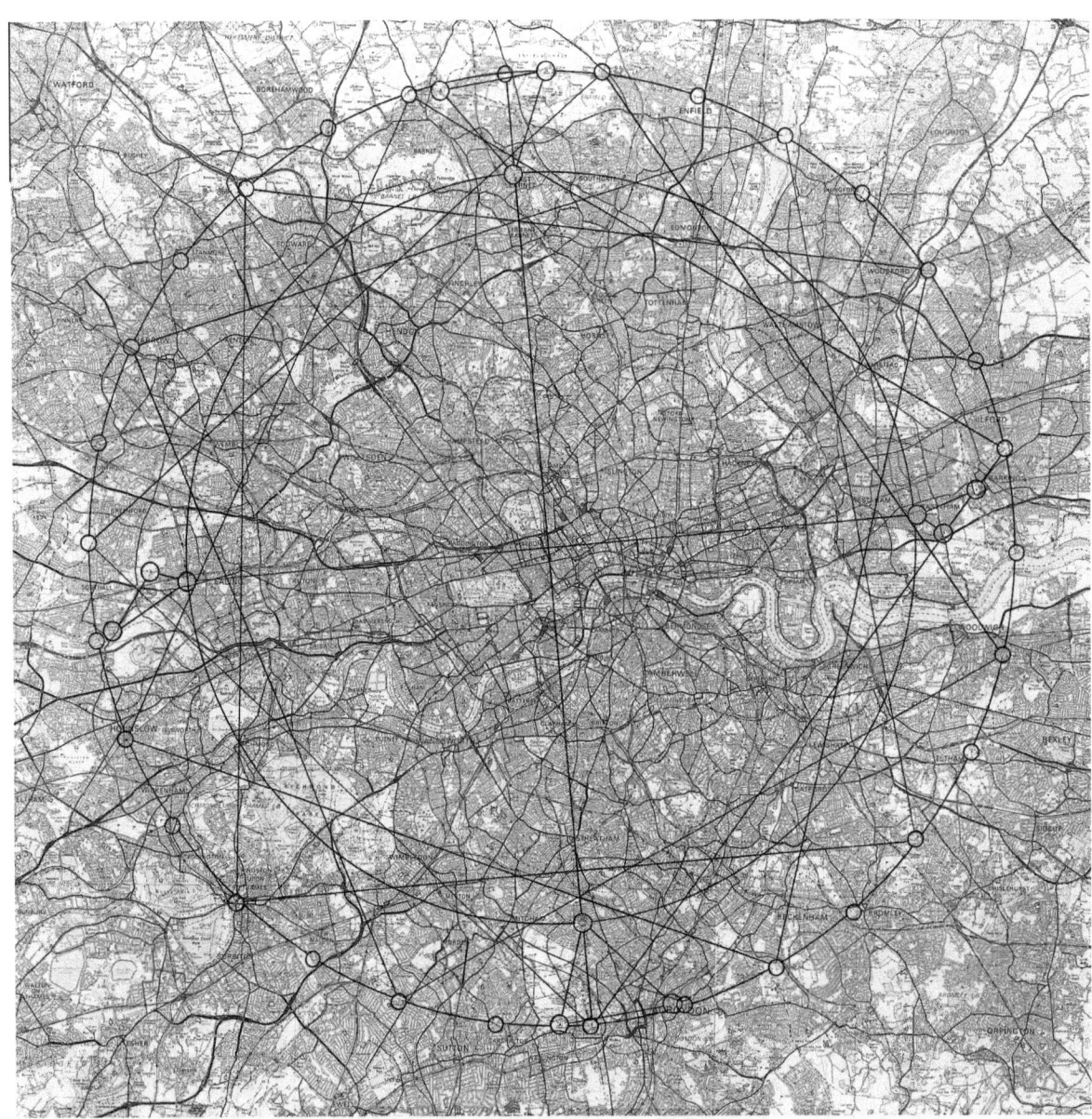

Illustration 90: Map of the 15-point star which shares every other point of the thirty . O.S. map © Crown Copyright. All rights reserved. License No. 100029558.

30-POINT STAR MARK POINTS

Clockwise from the most northerly point near Barnet:

1: Point in Camlet Way, Hadley Wood, near Barnet.
2: Hilltop point in private gardens near the junction of Hadley Rd and The Ridgeway, Enfield.
3: Point in playing fields at Enfield near the Queen Elizabeth Stadium.
4: King George's Reservoir, Enfield.
5: Field between Brook Rd and Whitehall Rd, Chingford.
6: Highfield Rd, Woodford Bridge.
7: Private gardens between Otley Dr. and Headley Drive, Gants hill.
8: Barking Park near the children's end of the boating lake.
9: The Creekmouth Flood Barrier.
10: Cage Lane Evangelical Church.
11: Eltham Warren Golf Course.
12: Court Farm Evangelical Church.
13: St. Martin's Park Bromley close to the War Memorial on the hilltop.
14: Crouch Oak Wood in the grounds of Bethlem Hospital Eden Park.
15: Ashburton Ave, Addiscombe.
16: Wasteland near a car park in Commerce Way, Waddon.
17: The Wrythe, formerly allotments off Wrythe Way.
18: St. Anthony's Hospital on the A24 in North Cheam.
19: L.S.E. sports ground opposite Balgazette Gdns in Windsor Ave, Kingston.
20: All Saints Church at Kingston-on-Thames where there has been a church for well over 1,000 years. Nearby is The Kings' Stone from which the town derives its name.
21: Strawberry Hill, Twickenham.
22: Whitton Rd near Hounslow Station.
23: The Glen, Norwood Green.
24: Pond near the Grand Union Canal between Greenford and Southall.
25: Hilltop at Dabb's Hill Rd, presumably this was Dabb's Hill, now topped by blocks of flats. The exact point is near a tree on a trianglular grass reservation.
26: Somerset Rd, Harrow.
27: Point on the grass central reservation in Woodlands Drive, Stanmore.
28: Point in fields east of the A41 near Elstree.
29: The grounds of Saffron Green School, Borehamwood.
30; Field north of Barnet Grid Sub Station off St. Alban's Rd, Barnet.

Firstly, despite the fact that the East Ham and Hanwell Triangles seemingly have no obvious relationship to St. Mary's East Barnet and Pollard's Hill, both sites feature amongst the new 15 and 30 stars' mark points because both stars are orientated on the North - South axis originally defined by St. Mary's East Barnet and Pollard's Hill. This is rather surprising since the 15 and 30 stars originate from the triangles of the East - West axis. The 30-point star actually links the East-West and North-South axes of the Earthstars patterns, although its North and South points are on the outer circle whilst the East and West points of the inner circle are junction points of its constructional lines. To link all four points on its own circle, it would have to be doubled to a sixty-point star.

These are not the only primary axes the latest stars share. Every single one of the five diagonals of the original five and ten pointed stars on the inner circle can all be extended directly to mark points of the thirty-pointed star on the outer circle. Their axes are shared. The mark points along the 15/30 stars' diagonals therefore include all the points of the ten-pointed star.

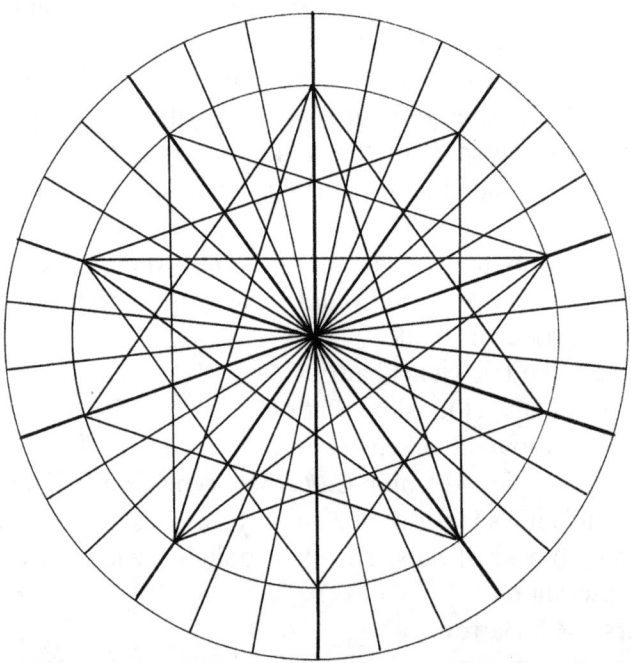

Illustration 91: Diagram showing how the all five axes of original ten-point star are shared with the thirty and extend to ten of its mark points. Like all the other Earthstars patterns, they fit together perfectly.

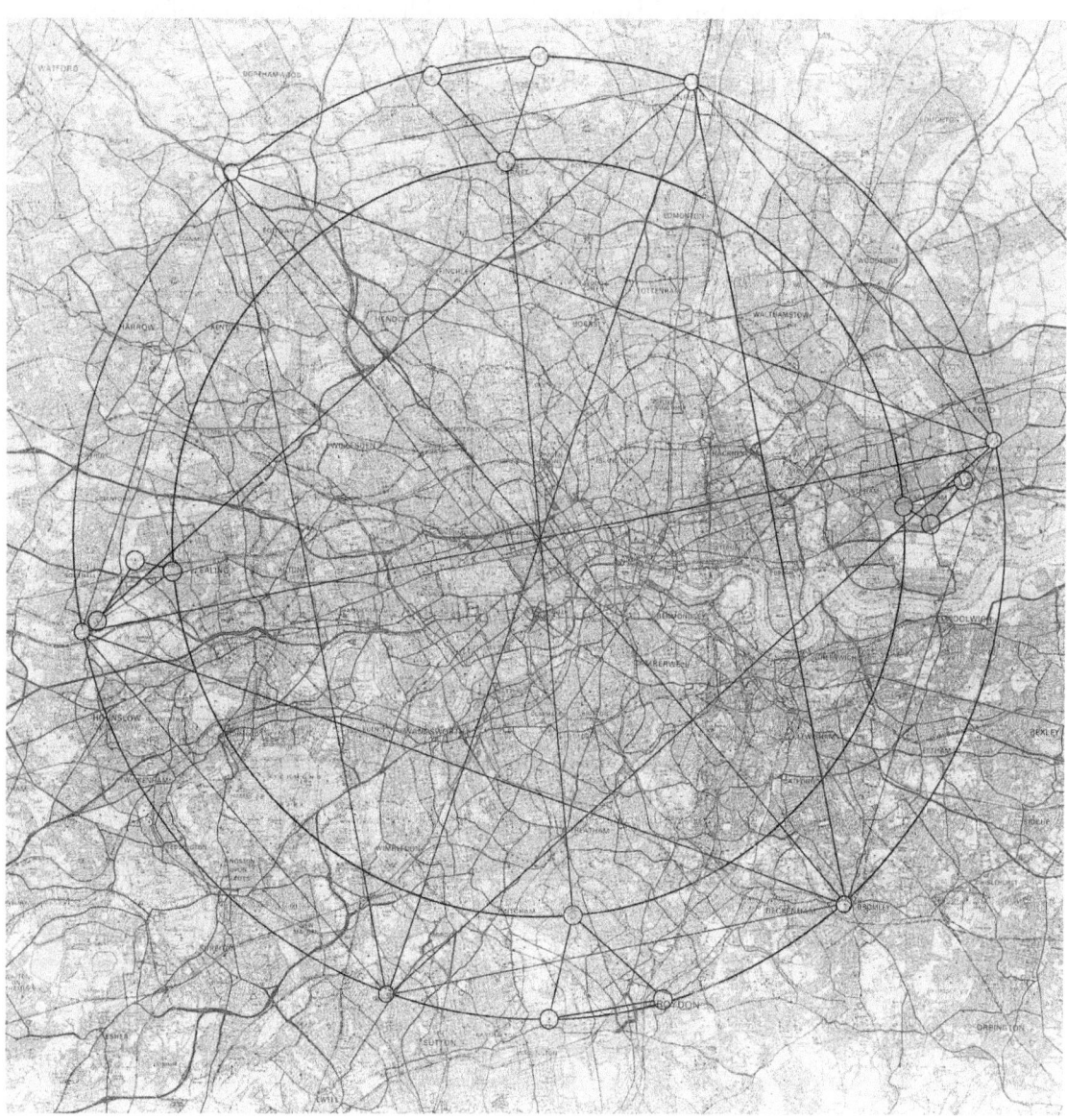

Illustration 92: The 30-point star and the eighteen-sided polyhedron share six points in common. They are linked by a perfect hexagram. O.S. map © Crown Copyright. All rights reserved. License No. 100029558.

Being a much larger figure than anything found previously, its tally of other important sites is also impressive. They include, the King Stone at Kingston Upon Thames, formerly the Coronation stone of the Saxon Kings, Westminster Abbey, Parliament Hill, the tumulus reputed to be Boudicca's resting place on Hampstead Heath, a hilltop with a startling statue of angels on a war memorial behind the Parish Church in Bromley, plus many other unique sites as well as numerous churches ancient and modern.

As we have seen, the 30-point star's axial lines confirm rather dramatically that it is an intrinsic part of the patterns already uncovered, but if further confirmation is required to convince you, it is readily forthcoming in the form of yet another fascinating set of connections. The thirty-point star shares six points in common with the eighteen-sided polyhedron which evolved from sides of the Barnet and Croydon triangles. As you'd expect by now, these are not six random points. They are evenly spaced in a perfect hexagram.

Curiously, it is offset from the majority of the other patterns by exactly 23 degrees, so that two of its points share an axis of the ten point star on the inner circle, creating an interesting fusion of pentagonal and hexagonal symmetry

Of course, the whole thirty-pointed star itself blends hexagonal and pentagonal symmetry is a startlingly harmonious manner. Apart from the pattern it forms within the Earthstars network, a thirty-point star can be drawn as five six-pointed stars, or six five-pointed stars.

The theme of pentagrams perfectly matched with hexagrams can be taken a stage further. If the thirty is drawn as five six-pointed stars, some of its hexagons on the outer circle would share common radial axes with the pentagonal patterns of the five and ten-point star on the inner circle.

Similarly, if the thirty is drawn as six five-point stars, some of its pentagrams on the outer circle would share common radial axes with a hexagram on the inner circle, aligned to the North-South axis of the existing patterns.

In sacred geometry, successfully combining the pentagram and the hexagram is regarded as something of an alchemical act and a marriage of opposites, symbolising the union of forces that create life.

In the context of Earthstars, it probably represents a fusion of polarities in the circuitry of the Earth Spirit.

This, therefore, is an enormously significant pattern.

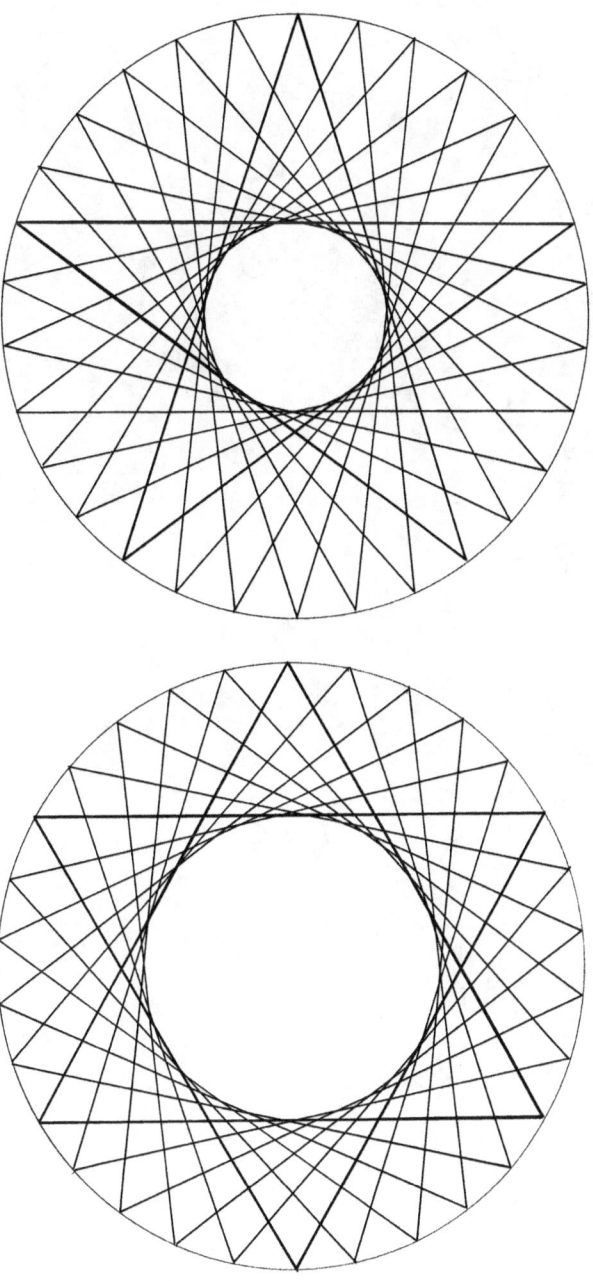

Illustrations 93 and 94: The 30-point star can be drawn as 6 x 5-point stars or 5 x 6-point stars, a perfect blend of the two symmetries.

London   City of Revelation

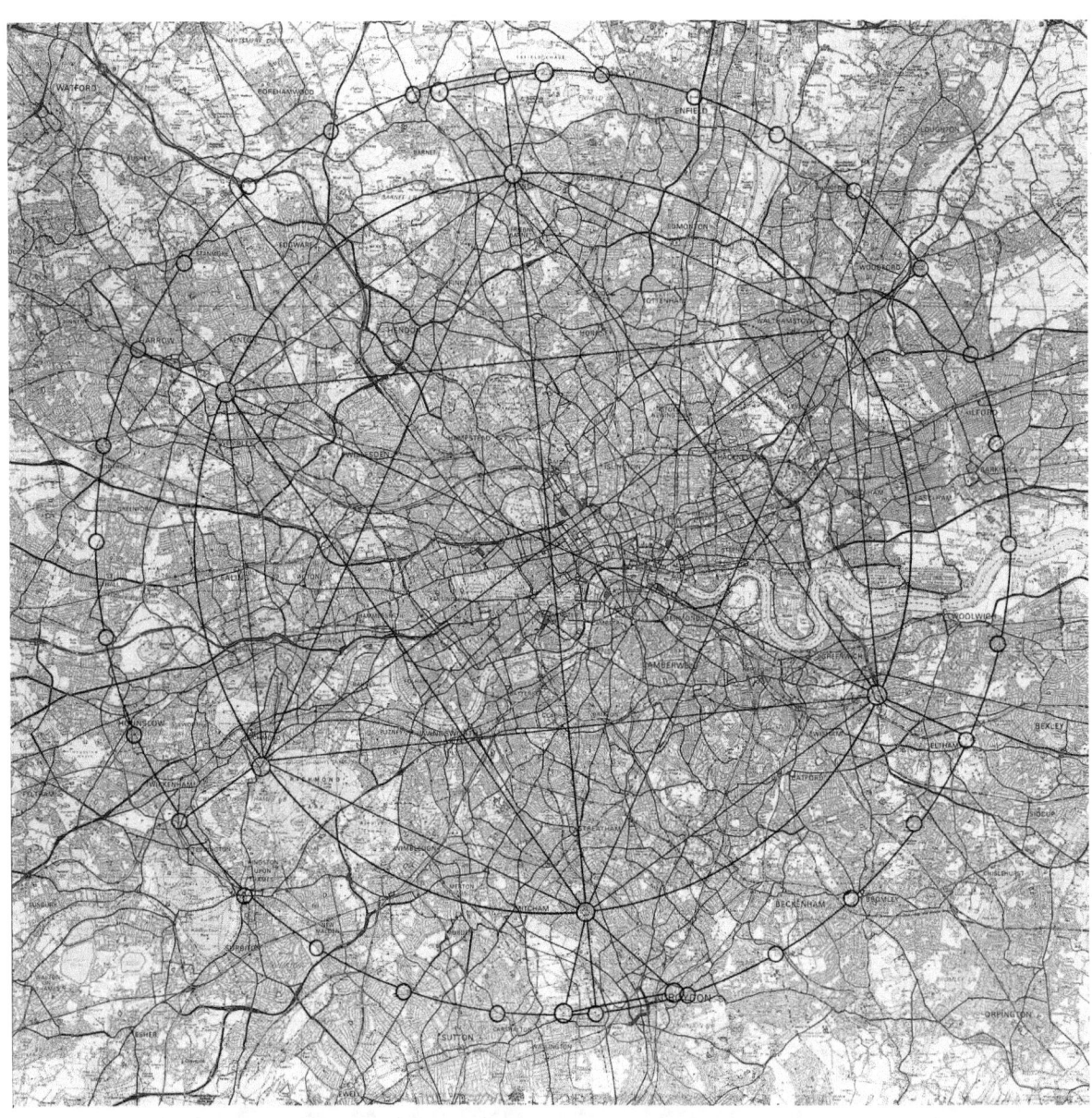

Illustration 95: The hexagonal star on the inner circle, identified by the axial lines of the 30-point star. Mark points clockwise from North; 1: St Mary's East Barnet.   2: Point in Epping Forest near Snaresbrook Rd. 3: St. John Fisher Catholic Church, Kidbrooke. 4: Pollard's Hill Norbury. 5: Richmond Hill. 6: Bryon Court Primary School, Kenton. O.S. map © Crown Copyright. All rights reserved. License No. 100029558.

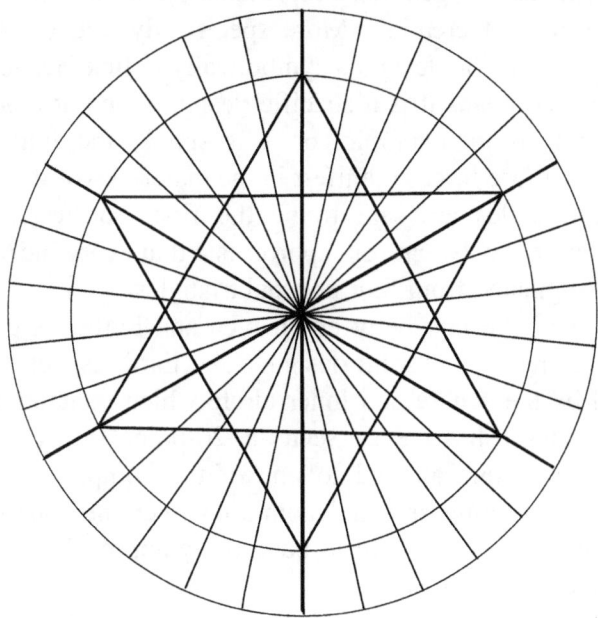

Illustration 96: As well as sharing the pentagonal axes of the original 5 and 10-point star on the inner circle, the 30-point star also shares hexagonal axes with the inner circle.

THE PATTERNS SO FAR

To sum up the discovery so far: from the simple beginnings of Barnet's equilateral triangle, we now have a grand total of sixteen patterns. Firstly, there are four triangles, from which all the other patterns evolve. From the Barnet Triangle, there's an eight, ten and twenty-pointed stars on the inner circle, plus a five and eighteen on the outer. From the Croydon Triangle, there's a pentagram on the inner circle and an eighteen on the outer. From the East Ham Triangle, there's a pentagram on the inner circle and a 15-point star on the outer. From the Hanwell Triangle, there's a pentagram on the inner circle and a thirty-point star on the outer. In addition, there are two hexagrams, one on the outer circle, one on the inner, both evolving from the thirty-pointed star.

Not one of these is an isolated pattern. Every single one interacts with one or more of the others to form a complex and beautiful mandala on the landscape. They are integral parts of the whole. In more ways than one. But what does this design represent ? If we are to believe the general symbolism, the sacred geometry represents the hidden forces of nature, or the divine forces of creation. More specifically, the basis of the overall design is a construction designed symbolically to link Heaven and Earth.

Either that means it is a specific design aimed at creating an area of sacred ground, as in a temple, or it is simply the fundamental pattern through which those forces manifest in the material world.

Since triangles are said to be the first manifestation of spirit in material form, it may be entirely appropriate that every individual pattern of this design originated from four unusual triangles.

As a symbolic link between Heaven and Earth, its geometry reveals some other extremely relevant designs. The Earth aspect of the formula is represented by the square and inner circle which contains the finest Earth energy mandala I have ever seen; a 20-point star composed of four pentagrams, each one aligned to one of the compass points. With each pentagram representing spiritual dominion over the four elements of the material world, they too prove to be entirely appropriate for their place in the grand scheme of things.

The fusion of heaven and Earth is symbolised on the outer circle of the heavens, by the thirty-point star, 30 points being a union of the Earth-based fives with the balanced solar sixes. As Ross Nicholls states in THE BOOK OF DRUIDRY; "5 is **the active material principle, six is the number of the sun and balance.**"

The number nine is associated with the feminine principle and the moon, so the 18 sided polyhedron may have lunar associations. It is linked to the 30 by another hexagonal balanced 6.

All of this, I know, is a little esoteric for the general reader and so may be difficult to comprehend and even harder to believe. There are times when I have to admit that I have found it hard to believe its implications myself.

What I have discovered here is not within the everyday experience of the vast majority. It was initiated by visions resembling The Virgin Mary and the entire discovery which has come to light as a result is not that much more tangible. To many people these things will be just unbelievable. Nevertheless, the Earthstars patterns are too big, too complex, their individual patterns too numerous and too well integrated as part of the composite whole to be anything other than a phenomenon with a foundation in some reality.

Far more evidence to support this is to come.

If it is unbelievable, I have to point out that it is also undeniable.

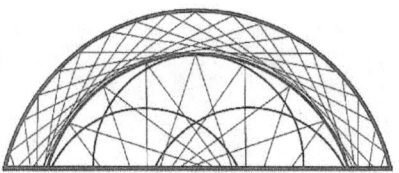

Chapter Nine
Practical Applications of Sacred Geometry

There is a wealth of evidence to suggest that sacred geometry, including identical constructions to those found in the Earthstars mandala, has been used in the design and construction of ritual sites for thousands of years. It is implicit within the sacred enclosures of megalithic stone circles, the impressive temples and pyramids of ancient Egypt, the classical temples of Greece and the Roman era, the great Gothic cathedrals of the middle ages and many modern structures whose design follows these traditional lines.

That this is ancient knowledge is clear from the work originated by Professor Alexander Thom and subsequently followed up by Keith Critchlow, John Michell, Nigel Pennick and others.

Thorm's extensive researches involving over 600 megalithic sites showed that while many rings are not actually true circles, their irregular shapes cannot be attributed to any ineptitude on the part of their creators. On the contrary, he found that the layout of many deceptively simple-looking stone rings incorporated extremely precise use of Pythagorean 3-4-5 triangles, more than a millennium before Pythagorus. But then Pythagorus is known to have acquired much of his learning from the Egyptians, whose temples also demonstrate an immense knowledge of arcane geometry, some of it pre-dating the European Megalithic monuments.

Many of the same patterns which occur in the Earthstars design underlie Egyptian monuments through a shared basis in sacred geometry. However, there are far more direct and momentous connections. Astonishingly, the Great Pyramid of Khufu at Giza has direct associations with the same basic framework as the Earthstars patterns, the squared circle.

I am indebted to John Michell for this information which I gleaned from his book, CITY OF REVELATION. If a cross section of the Great Pyramid is superimposed on the Earthstars design, with the width of the pyramid's base corresponding to the diameter of the Circle of the Earth Spirit, then the tip of the pyramid would touch the outer circle, the Circle of the Heavens.

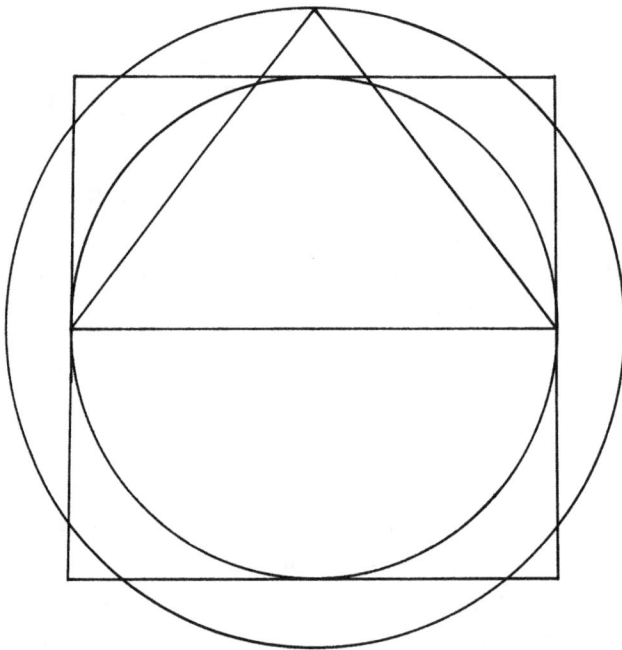

Illustration 97: If the squared circle geometry is three dimensional and the Great Pyramid stands on the Square of the Earth, its apex would touch the Circle of the Heavens.

To appreciate the full significance of this relationship, think of the squared circle as both a ground plan and three dimensional design. If the pyramid's base sits neatly on the square of the Earth, its apex would touch the Circle of the Heavens. Like the squared circle diagram, the Great Pyramid is a construction designed to link heaven and Earth. In Graham Phillips' book ACT OF GOD, he states:

> **"Modern commentators have attributed the pyramid with astrological, mystical or even alien significance; the only thing of which we can be certain, however, is that they are seen as a link between heaven and earth. The name pyramid is a Greek word and was the term they first used when they encountered these monuments. The Egyptian word for these structures was MER, meaning PLACE OF ASCENSION."**

The Great Pyramid isn't the only impressive ancient monument which utilises the Earthstars geometry in its ground plan. John Michell's work demonstrates that Britain's most famous and enigmatic temple, Stonehenge, also incorporates the squared circle design. This encouraged me to overlay the Earthstars design upon John's diagram of Stonehenge. Astonishingly, I found my patterns fitted as accurately as his.

The two concentric Earthstars circles matched the concentric bluestone and sarsen rings perfectly. Even more surprisingly, every point of the thirty-point star on London's outer circle matched a stone of the sarsen ring. Up until that point, I had no idea how many stones were in that circle. It simply hadn't been important before. Now it was startlingly so. There were thirty upright stones, plus thirty horizontal lintels on top of them.

Illustration 98: The concentric circles of Stonehenge's sarsen and bluestone rings are identical in proportion to London's Earthstars circles and the squared circle construction (illustration reproduced from John Michell's work with permission).

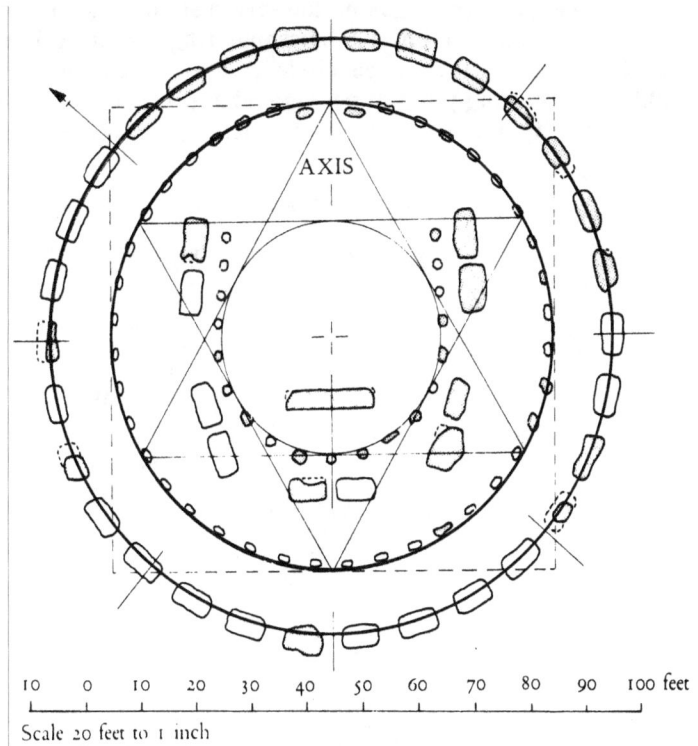

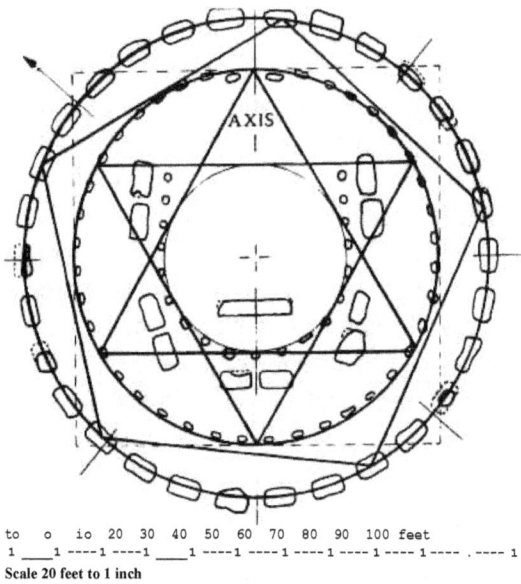

Illustration 99 and 100: Pentagonal and hexagonal symmetry are incorporated in the layout of Stonehenge. A pentagon on the sarsen stone ring circumscribes the bluestone ring. A hexagram on the bluestone ring circumscribes the inner sanctum. A pentagram on the sarsen circle creates a pentagon on the inner sanctum making it a meeting point of pentagonal and hexagonal symmetry. The principles apply equally to the Earthstars geometry (see pages 104-106).

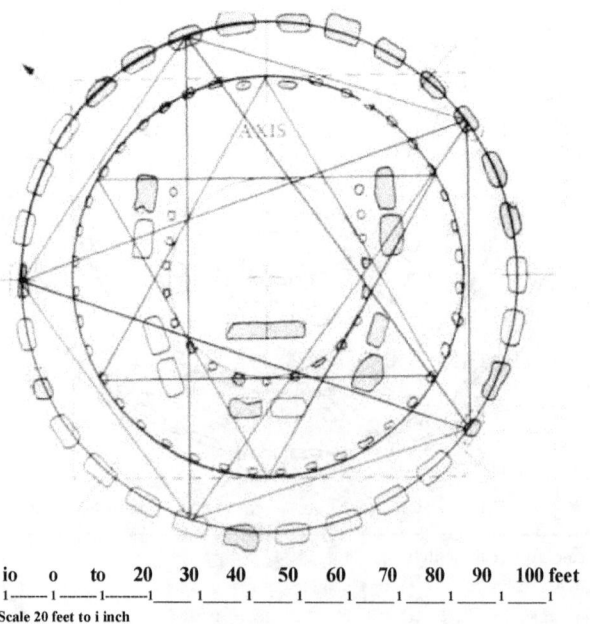

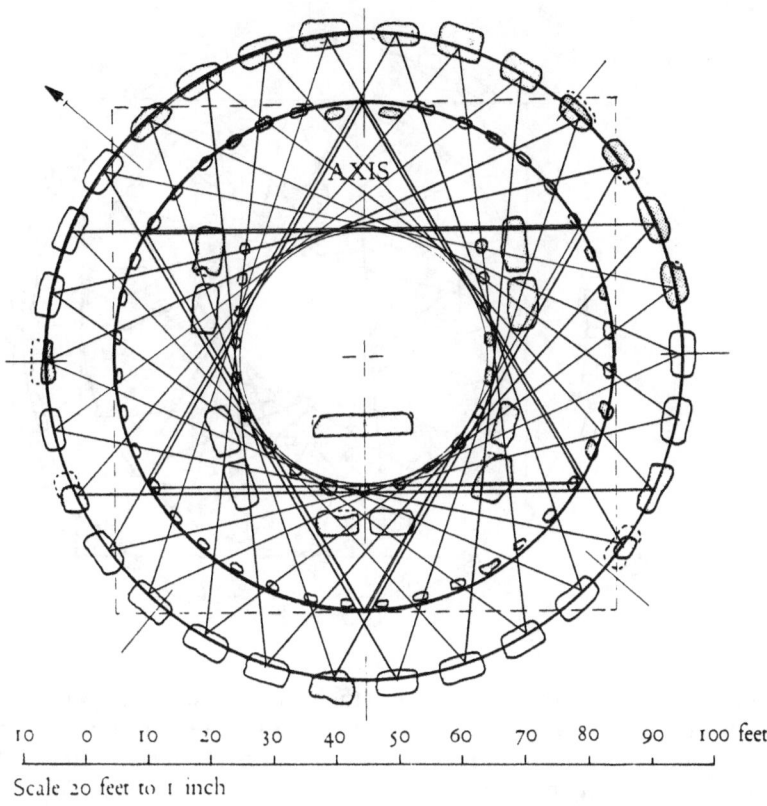

Illustration 101: London's 30-point Earthstar fits Stonehenge's ring of 30 sarsen stones perfectly and even incorporates the hexagram on the inner circle, as well as enclosing the inner sanctum of the bluestone horseshoe.

Nor had I realised there were forty stones in the bluestone ring. London's twenty-point star on the inner circle fitted very nicely onto it, with the twenty points highlighting alternate stones.

This is doubly significant since the twenty and thirty-point stars can be super-imposed on Stonehenge together on their appropriate circles. Given the complexity of these patterns, that can hardly be a coincidence.

This impression is confirmed by yet another correspondence. The Earthstars eight-point star which sits upon the same circle as the 20-point star in London's patterns overlays on the very same circle at Stonehenge, the bluestone ring.

When the 20 and 8-point stars are overlaid, the twenty super-imposes a pentagram upon four of the eight's points. Stonehenge's forty stone circle could be based on an eight-point star design with a pentagram overlaid on all eight points; eight, five-point stars. Equally, it could be drawn as five, eight-pointed stars. Either way, it is an

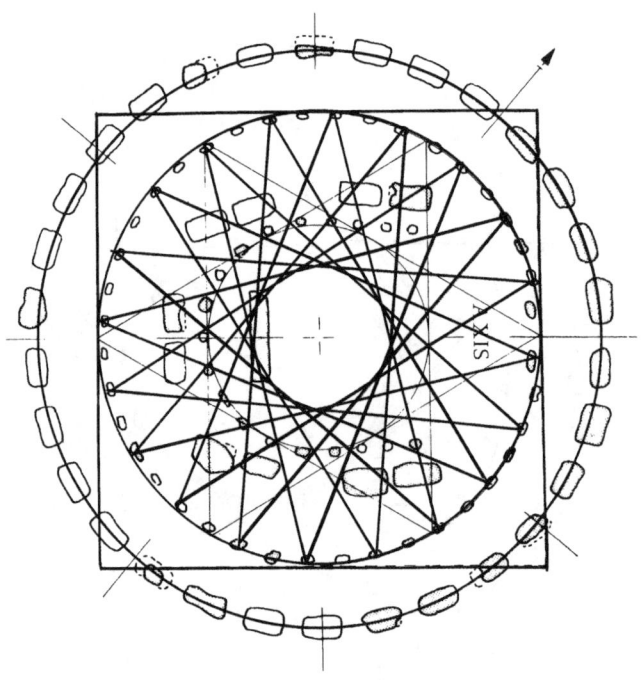

Illustration 102: The 20-point Earthstar also finds a perfect match at Stonehenge. Not only does it overlay neatly onto the bluestone ring which has 40 stones (a double twenty) it is on the correct circle relative to the thirty. The geometry of Stonehenge is virtually identical to the London Earthstars' patterns.

intriguing combination of pentagonal and octagonal symmetry Other aspects of the Earthstars' geometry highlight the relative diameters of the concentric circles. This was most obvious with the 30-point star whose chords delineate the inner sanctum of the bluestone horseshoe at the centre of the monument.

Once again, the relative measures turned out to be in holistic proportions, reflecting cosmic measures. Stonehenge's 108 ft diameter (108, lunar radius number) sarsen ring would fit into London's Earthstar Patterns precisely 792 times. Not roughly. Precisely 792 is an Earth diameter number.

One accurate correspondence between the Earthstars geometry and Britain's most famous ancient monument would have been impressive enough. To find all these was incredible. A similarity between the patterns might have been expected through their common basis in sacred geometry, but these patterns aren't just similar. Large parts of them are identical. The fundamental framework of Stonehenge and London's Earthstars is a perfect match and not just one, but five of the star patterns can be overlaid meaningfully, including the two most complex which fit perfectly.

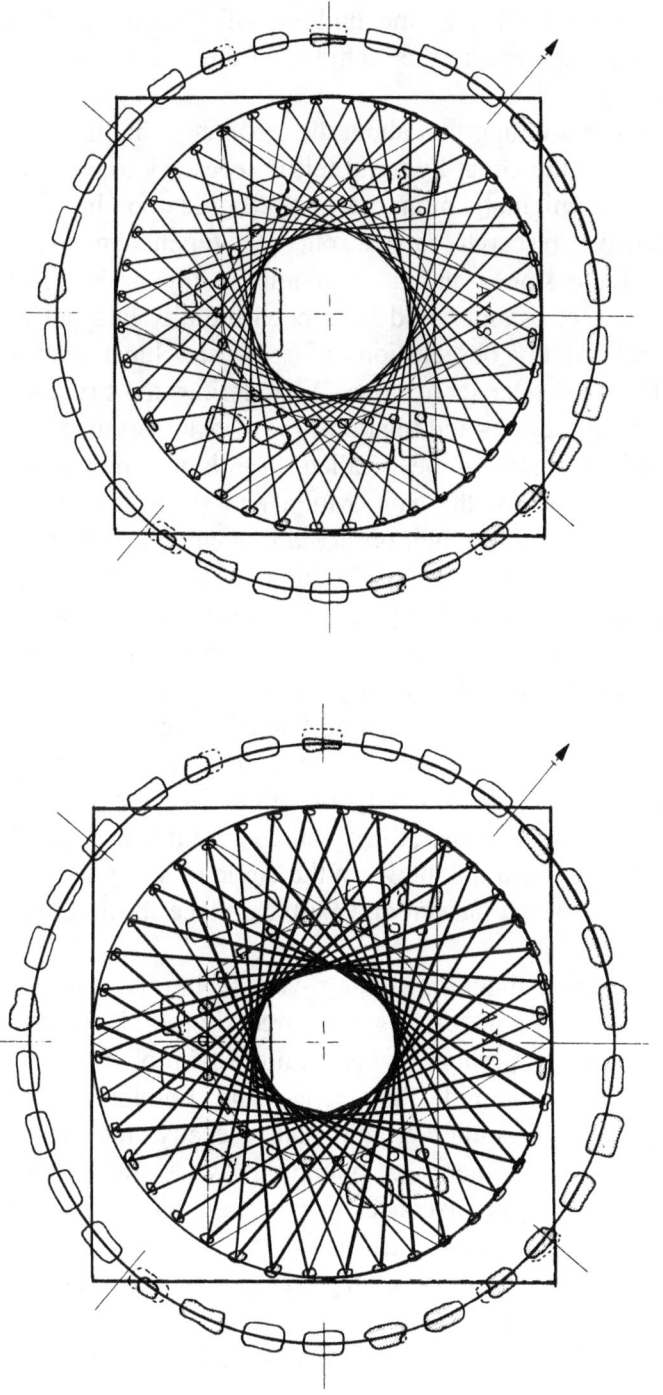

Illustrations 103 and 104: Stonehenge's ring of 40 bluestones could be based on eight five-pointed stars, or five eight-pointed stars.

This leads to the inescapable conclusion that the Earthstars design must have been known to the builders of Stonehenge more than three thousand years ago and to the architect of the Great Pyramid considerably earlier.

What was so important about this geometry that the architects of both Stonehenge and the Great Pyramid felt it necessary to base two of the most impressive and enigmatic monuments of all time upon it ?

Extensive research carried out at Stonehenge indicates that the positions of the stones and their geometric layout is directly related to astronomical factors and could have been initially laid out as the result of detailed and lengthy observations of solar and lunar cycles. It has been described as a calendar set in stone. What defines the structure of a calendar is the movement, cycles and relationships of planetary bodies.

Stonehenge has also been called, in folklore, the Giant's dance. If the giants in question are the Sun, Moon and other planetary bodies, their dance could well refer to their observed movements around the night sky as they dance across the heavens. In this context then the design can be understood as a layout of the cycles of the powers that turn the universe.

This, of course, is merely another way of expressing what the symbolism of the sacred geometry has already told us, but at the risk of being repetitive, here it is again. Sacred geometry embodies laws of proportion which underlie the physical world. As such, its designs may be taken to represent, symbolically or otherwise, the formative forces of creation. Within this context, the squared circle fulfils a specific function. It is designed to create a link between the circle of spirit and the square of the Earth; a link between heaven and Earth; a place in direct contact with the heavens.

As a simple diagram, it may well be purely symbolic. But as the fundamental design underlying the Great Pyramid, Stonehenge and the Earthstars mandala, perhaps this is not mere symbolism.

The Great pyramid, Stonehenge and London's Earthstar patterns could be based on similar ground plans because the designs themselves embody and focus otherwise intangible forces and enable them to be applied, consciously or subconsciously at these locations.

Are they all "places of ascension" ?

Are they really some kind of stargates which facilitate shamanic soul-travel beyond the Earth's realms ?

Do they literally connect us to the stars ?

BUILDING IN THE SPIRITUAL DIMENSIONS

To understand more fully the nature, function and purpose these patterns represent, we have to look into the practical applications of sacred geometry and its widespread use in the design of temples, churches and cathedrals throughout history.

Architects of all ages and cultures seem to have employed the general principles of sacred geometry to provide a hidden foundation for the external form of all kinds of religious monuments and structures. Egyptian, Roman and Greek temples, with thousands of years, as well as thousands of miles, between them all incorporate Plato's Golden Mean in the proportions of their design.

The great Abbeys and Cathedrals of Europe, along with Saxon, Norman and Gothic churches can all be shown to exhibit various examples of sacred geometry as an esoteric aspect of their construction. Even the Globe theatre was based upon a ground plan of sacred geometry similar to that used in temples.

Illustration 105: An Egyptian rock tomb at Mira clearly based on pentagonal/decagonal golden mean principles.

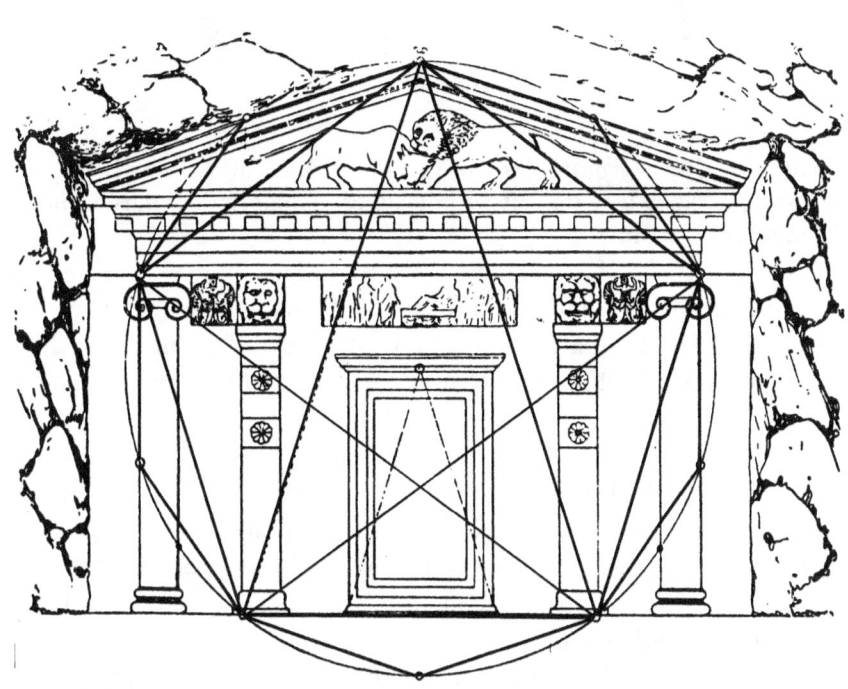

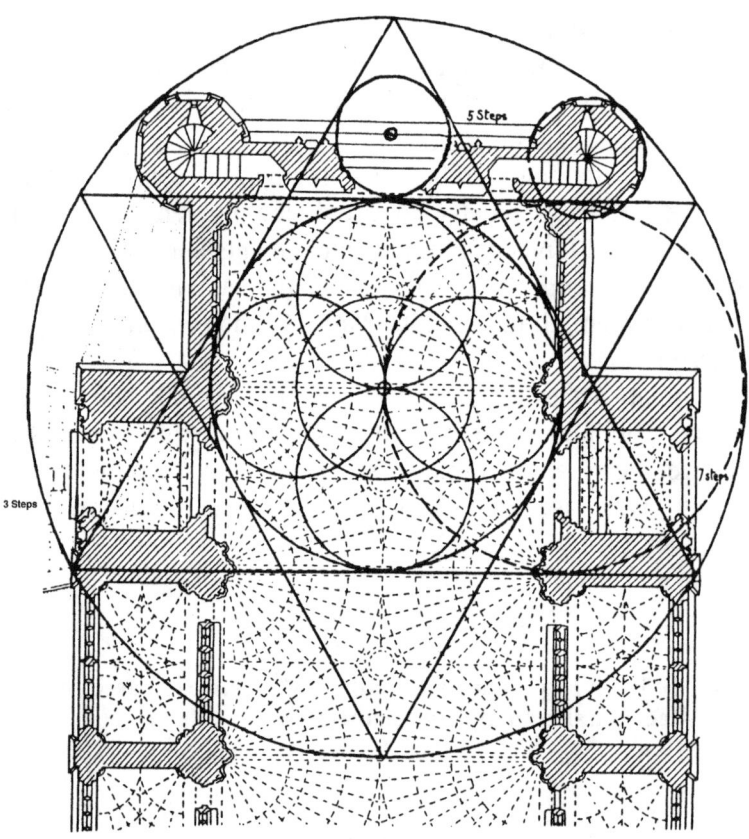

Illustration 106: Groundplan of the West End of King's College Chapel Cambridge, showing that its underlying geometry is based on the Ad Triangulum system.

According to Keith Critchlow and Nigel Pennick, two of the most respected authorities on the subject, the masons and architects of the middle ages employed two distinct systems of sacred geometry as the basis of their constructions; Ad Quadratum and Ad Triangulum.

In his extremely well-researched book, SACRED GEOMETRY, SYMBOLISM AND PURPOSE IN RELIGIOUS STRUCTURES, Nigel Pennick states that the older of the two systems was Ad Quadratum, which evolved from an earlier pagan system related to the eight directions. This must have been a western equivalent of Feng Shui whose basis is the eight directions of the Pa Kua.

Basically, Ad Quadratum uses a template of interlocking squares

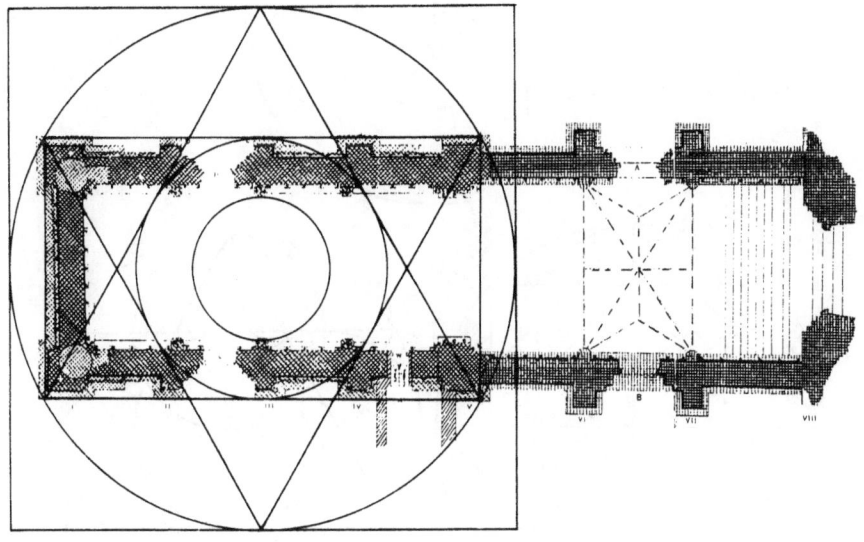

Illustration 107: The hexagonal (Ad Triangulum) basis of The Lady Chapel at Glastonbury Abbey.

as a framework upon which the actual design of the building could be laid out. Ad Triangulum, he explains, was a similar design system based on equilateral triangles and hexagrams.

Examples of these are relatively common and include Westminster Cathedral, Kings' College Chapel in Cambridge, Glastonbury Abbey and even Guildford Cathedral which was built as recently as 1936 upon an Ad Triangulum plan.

The predominance of five-point stars in the Earthstars mandala made me far more interested in sacred geometry based upon pentagonal patterns. Several examples have come to my attention including a Gothic Standard Plan utilising a ten-point star design of two interlocking pentagrams. Curiously, I have not yet found any direct reference to this system. I attributed this absence of information to the possibility that mysteries of the pentagram and Golden Mean meant that it may have been regarded as somehow even more esoteric than the others. That the three systems are related can be seen from their foundational geometry.

The four-fold symmetry which provides the basis of Ad Quadratum can also be shown to be a framework upon which Ad Triangulum may be constructed.

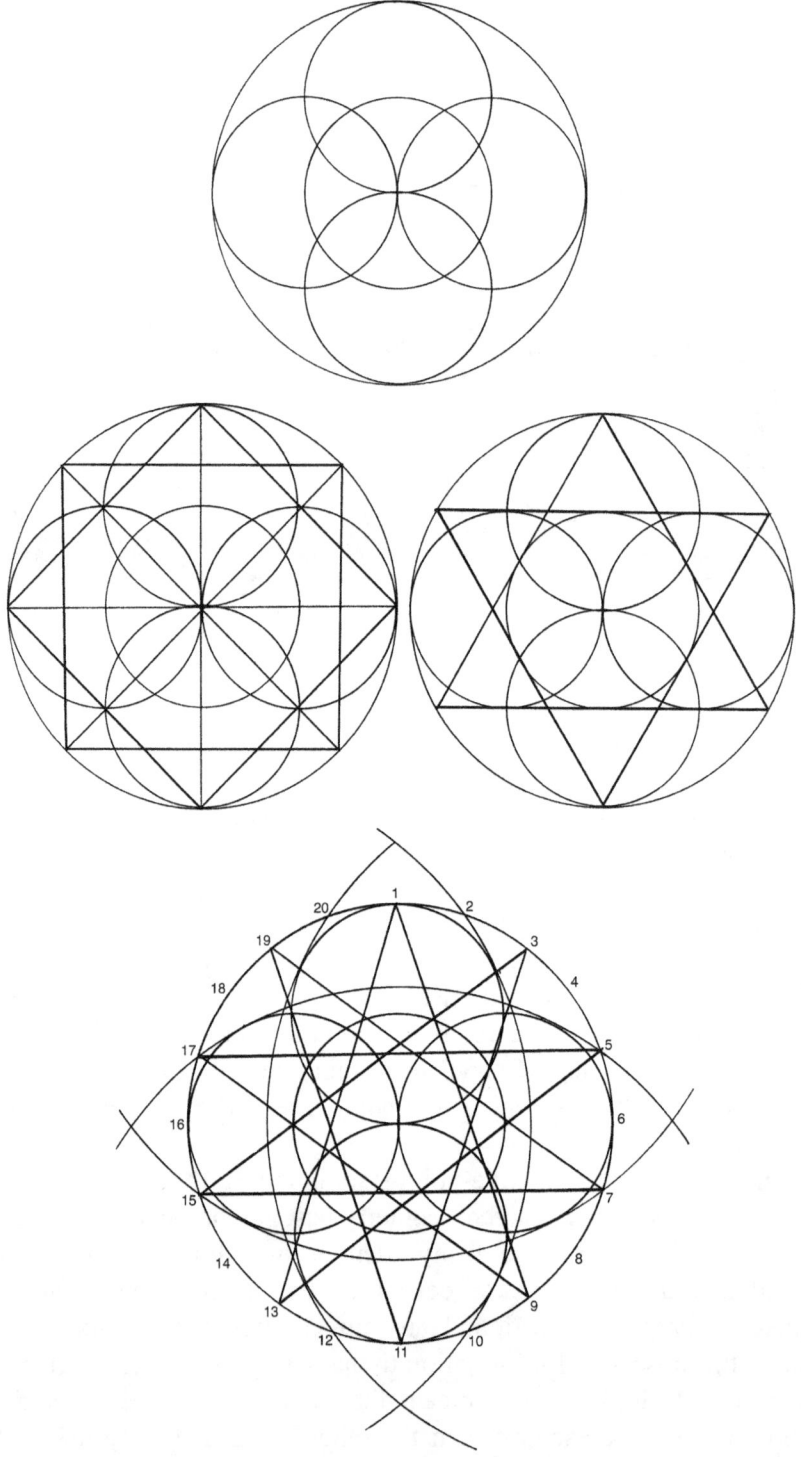

Illustrations 108 a, b, c, and d: The four-fold symmetry which provides the basis of Ad Quadratum can also be shown to be a framework upon which Ad Triangulum may be constructed. It also forms the basis of a pentagonal construction of the squared circle and the 20-point star of London's Earthstars.

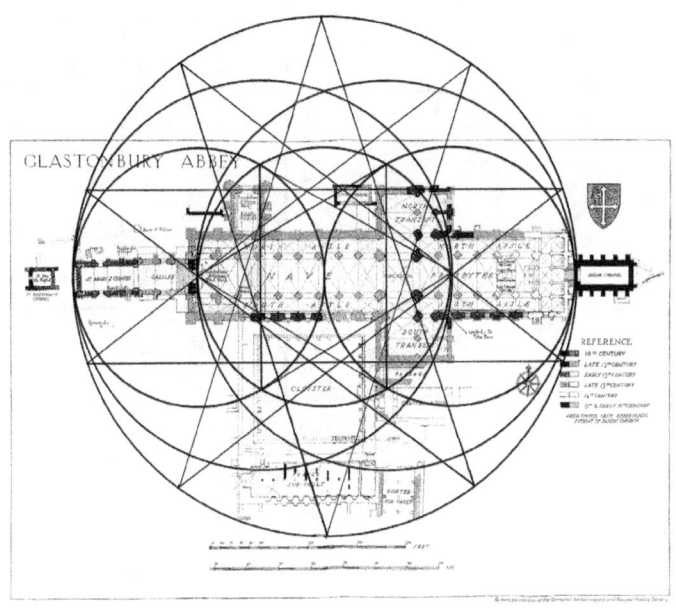

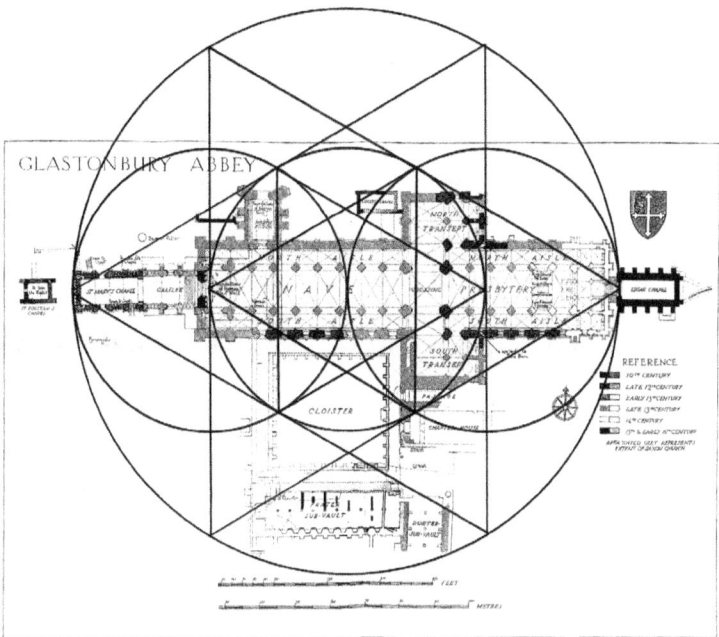

Illustration 109: Glastonbury Abbey displays an Ad Triangulum basis, as well as links to the 10-point star of the Gothic Standard Plan.

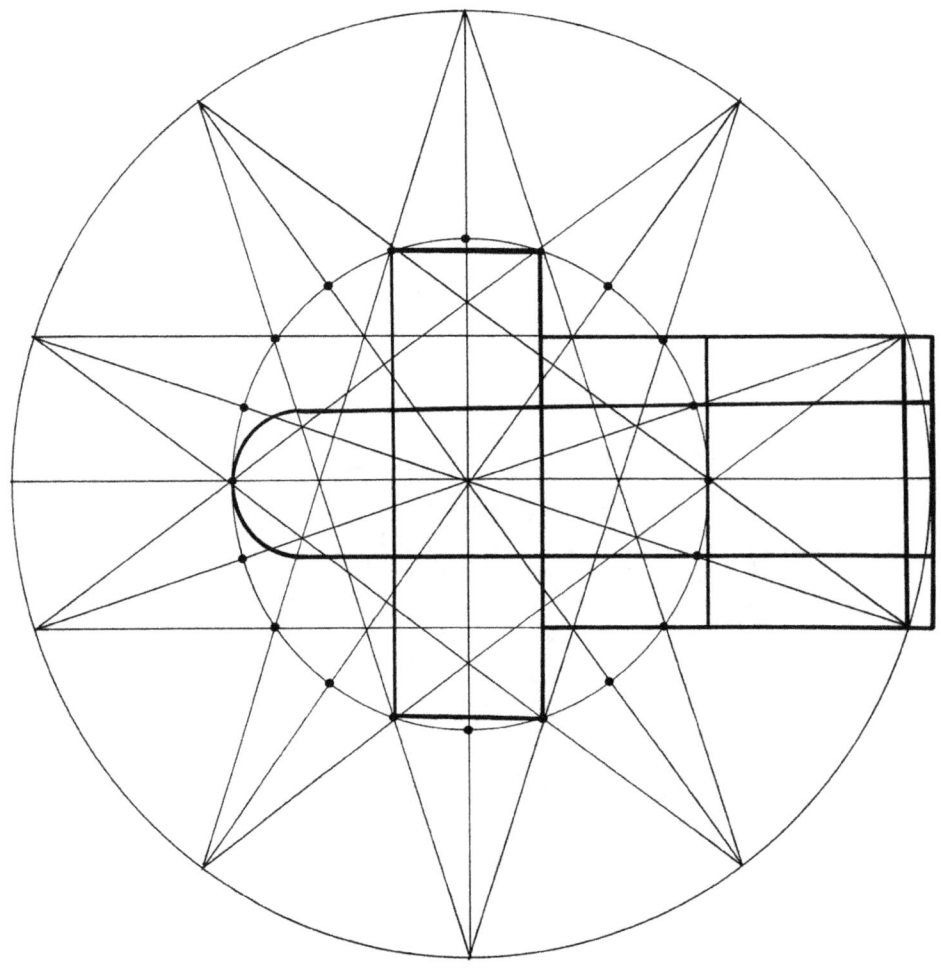

Illustration 110: Gothic Standard Plan based upon a ten-point star identical to one of London's Earthstar patterns. If you count around the points of the inner circle, you'll find that the geometry of the apse, nave and transept are defined by a 20-point star.

City of Revelation

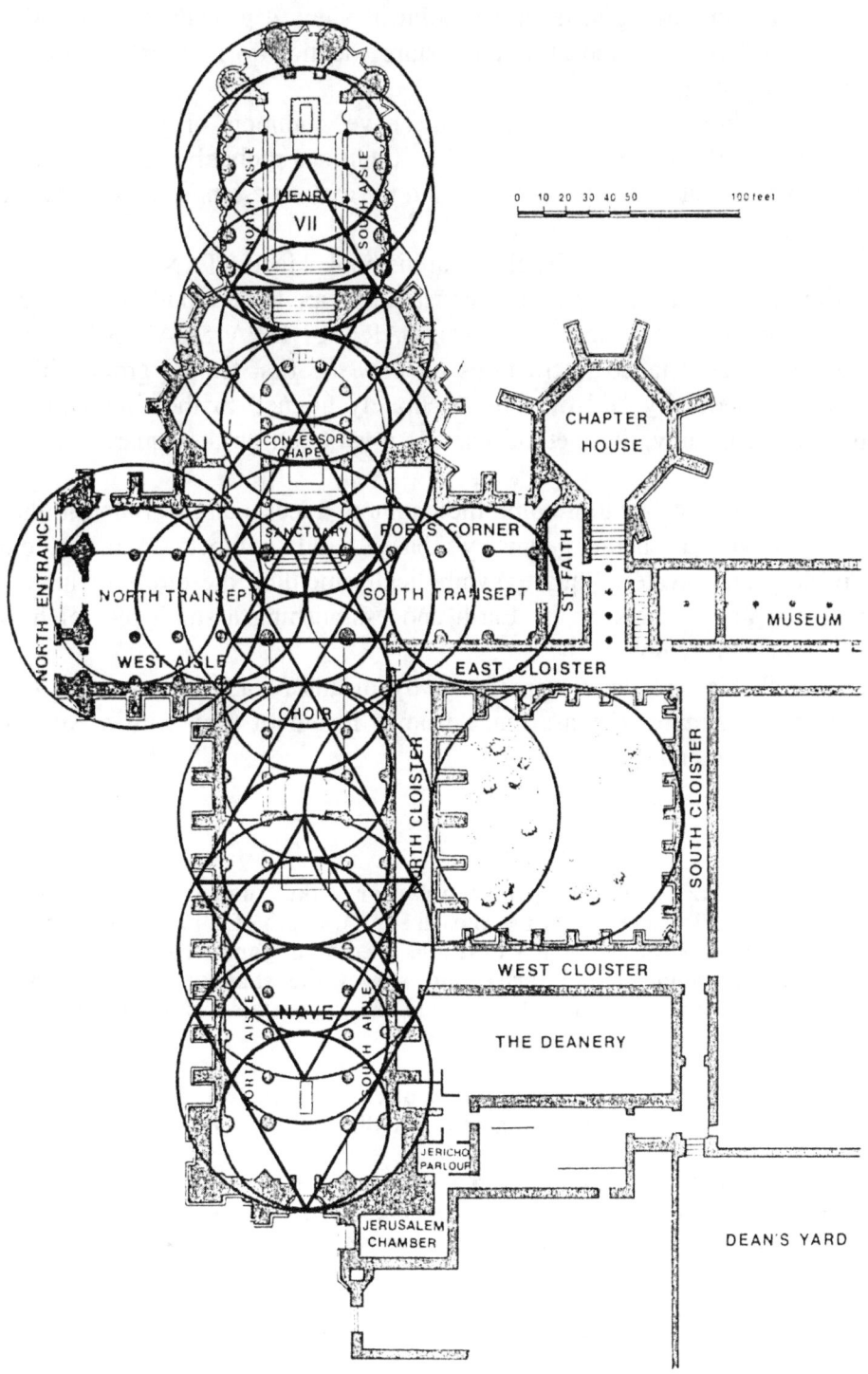

Illustration 111: Westminster Abbey ground-plan with an incomplete and rather poor geometric overlay.

More interestingly, from my point of view, it provides a foundation from which five, ten and 20-pointed stars, identical to London's Earthstars geometry, naturally evolve.

All three systems undoubtedly have common origins, as Nigel Pennick rightly suggests, in much older traditions probably stemming from our most ancient shamanic practices and the earliest establishment of sacred spaces.

According to Keith Critchlow in his book TIME STANDS STILL the earliest records relating to temple construction are to be found in an ancient Hindu Manuscript called the MANASARA SHILPA SHASTRA which actually gives detailed instructions on how the first temple ground plans were geometrically laid out and accurately aligned to the four cardinal points. Predictably, the method itself is a variation on the squared circle's construction.

It begins at a central point, the gnomon, traces a circle around it, then uses the vesica, sacred geometry's basic building block, to make four divisions and create a square, symbolically melding the circle of heaven with the four corners of the Earth and establishing a small plot of holy ground for the temple.

All this was done, apparently, by means of a device known as a trammel, a simple compass made from a length of rope with a wooden stake at either end.

Illustration 112: First an upright pillar or stake known as a gnomon is erected to mark the centre of the chosen site. A length of rope twice as long as the gnomon was then attached to it and used as a trammel to trace a circle around the gnomon. At sunrise and sunset the shadow of the gnomon intersects the circumference of the circle at two points which could be joined to give an accurate east - west axis.

Illustration 113: This east-west alignment can be transferred to the centre of the circle simply by creating a parallel line through the centre.

Illustration 114: Using the east and west points as centres and with a trammel equal to the diameter of the circle, two arcs are inscribed around the circle. Joining their points of intersection provides an accurate north-south axis.

Illustration 115: The process is repeated from the north and south points to form a second vesica around the circle.

Illustration 116: The circle of heaven "squared" to form the final temple groundplan and basic Ad Quadratum design.

London   City of Revelation

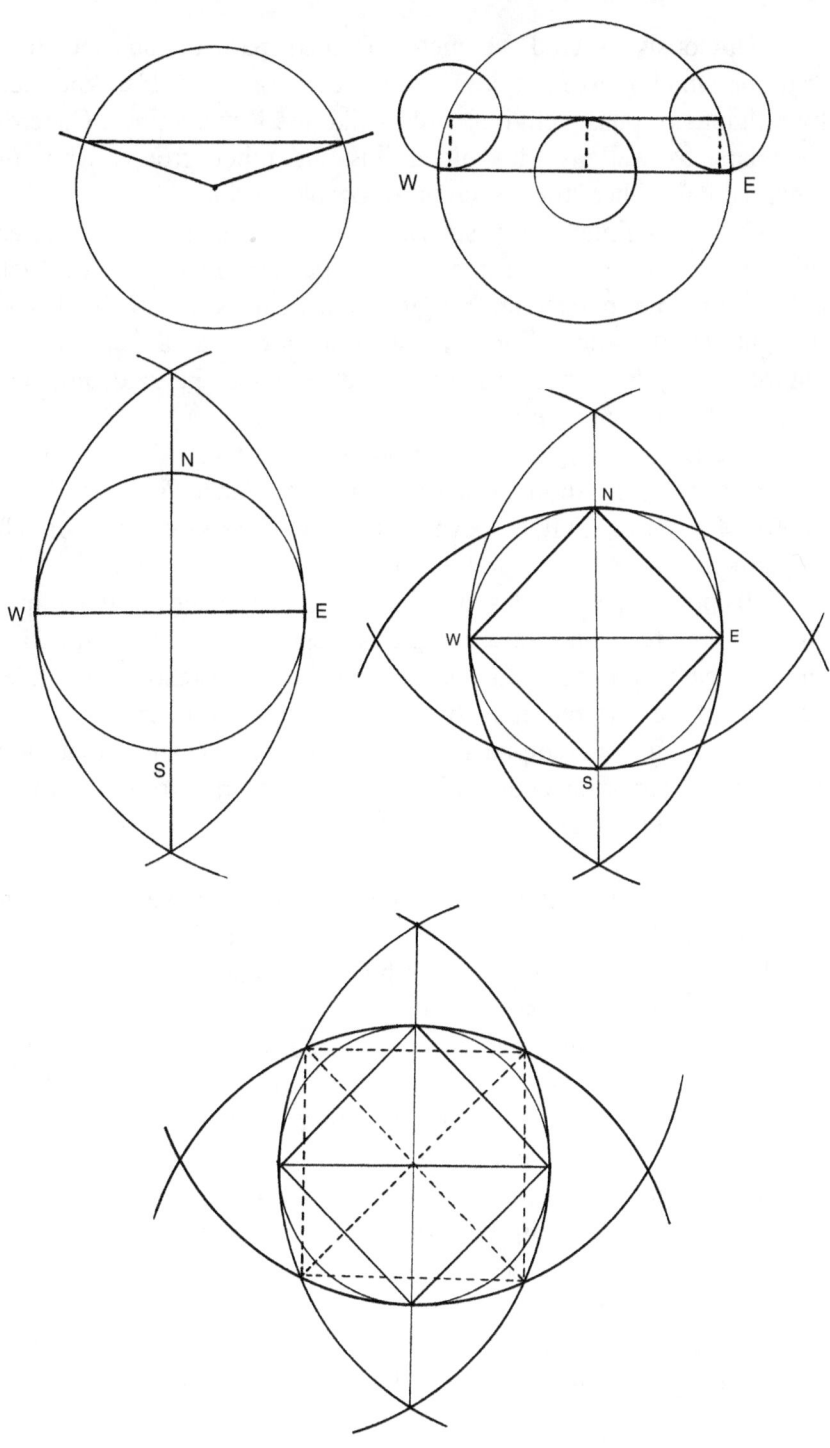

Obviously, sacred geometry formed part of an ancient wisdom tradition which must have survived the centuries as hidden knowledge held by religious or quasi-religious orders like the Benedictines, Cistercians and Templars, as well as Masonic guilds and other groups who may have regarded it as either "trade secrets" or occult wisdom.

It was definitely an esoteric aspect of the orthodox church in the middle ages, but how far the knowledge and awareness of such things extended within mainstream religion is unclear. Not very far, I would have thought. In all probability, only an elite group at a very high level of authority may have been involved and, it has to be said, not necessarily confined to any single religious order.

Sacred geometry is not just one isolated area of understanding. It is part of a complete, holistic system. As Joy Hancox's researches in THE BYROM COLLECTION reveal, it was clearly a part of Hermetic, Rosicrucian and Masonic philosophies.

Its presence in the design of our ancestors' places of worship suggests that the architects and master masons responsible were well aware of the importance of proportional harmony and holistic resonance. Indeed, that was one of the very reasons why they used this symmetry.

Working from templates of sacred geometry ensured that the builders were using proportions compatible with the natural world and the universe around us; thus man's additions to the landscape followed the same principles as the creators.

Just as these patterns may be construed as the underlying divine influence in the natural world, there is good reason to believe they were similarly understood in their architectural context; as a metaphysical pattern determining physical form.

Any stone, bricks and mortar erected upon a plan derived from the appropriate holy template would have deemed to have been built upon firm foundations in the spiritual realms as well as upon Terra Firma.

Since those patterns were in a harmonic holistic relationship with the structure of the planet, they ensured that the building was integrated with the cosmos, resonant with the primary spiritual powers, and could act as a vessel through which those powers could flow.

In short, a perfect place for worship and spiritual communion with any concept of the creator, whether it be in the form of the classical gods or goddesses of antiquity, Islam's Allah, or a Christian concept of God The Father.

THE GEOMETRY OF THE SOUL

Modern architects rarely incorporate this spiritual dimension into their work. The materialistic perceive the world around them as inanimate. They live in a dead world and their additions to it reflect that paradigm. Consequently when we describe some of their creations as soul-less, we are not making a subjective statement about the atmosphere of the place, we are being quite literal. They are soul-less. The architectural equivalent of zombies. They create dead buildings and a dead environment.

Hardly suprising then that they have much the same effect on their occupants; the opposite of making your spirit soar. Instead, it plummets.

By contrast, the masons who raised the great religious houses and monuments of the past would have regarded their sacred geometry as of paramount importance, the most essential part of the design. In their eyes, no religious establishment could possibly be expected to fulfil its primary function as a centre of divine power without it. A building without a soul could never be sacred or uplifting.

More important, when constructing a House of God, the soul of the structure would be analogous to the Holy Spirit and the correct proportions and dimensions ensured its active presence.

An impressive side effect of sacred geometry is that many of these structures possess extraordinary acoustic properties. A result of the inter-related harmonic structure of both music and geometry.

Music is harmony of structure in time. Geometry is harmony of structure in space. Sacred number underlies both.

Certain types of music were designed to be sacred geometry you can hear, so it is little wonder the buildings resonate with it.

Sacred geometry has been utilised in the stone age by those whose gods are almost forgotten, by the ancient Egyptian followers of Ra, Isis, Osiris and Horus, by the pagans of the classical Greek and Roman Empires, in Islamic mosques and in Christian Churches.

It is not exclusive to one particular sect or religion. It transcends the disparities of religious belief.

Like the forces of creation or Universal Life Force it represents, it is sacred to all people.

The web of life it weaves throughout the Earth and Cosmos is the hidden unity which links all religions to a common origin and experience of the divine.

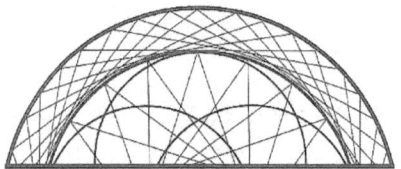

Chapter Ten
The Web of Life

If sacred geometry can be synonymous with the soul of a building, what do the same patterns represent in the landscape ?

The soul of the land ?

Out of context, you could be forgiven for dismissing this suggestion as completely nuts. However, when related to the Earthstars mandala and the unusual circumstances associated with its discovery, it actually make a lot of sense.

Traditionally, the spirit of the land of Britain is portrayed as a goddess; Brigid, Bride or Brittania. She is the Genius Loci of these isles. In turn, she is one of the aspects of the Earth Goddess.

Since apparitions fitting the description of these ancient deities (some also resembling the Virgin Mary) had led to this discovery, one might conclude that these phenomena could be an anthropomorphic manifestation of the Earth Spirit, the planetary life-force. The alchemist Robert Fludd speaks of mother nature in these terms:

" She is not a goddess, but the proximate minister of God at whose behest she governs the sub-celestial worlds. She is the soul of the world."

We humans, of course, are part of the World Soul, or at least our souls are, just as the world soul is part of the Universal Soul. All things are connected.

In confirmation of my own feelings on this matter, Fludd considered the life-force, nature, the Anima Mundi, the World Soul and the spirit of the Earth were all connected, if not simply different terms for the same thing.

Elusive though it may be, the life-force has been known throughout history by many names. Every mystic tradition tells us that all things are connected. What connects them is an invisible web that is both spirit and energy, the universal life force.

To an Indian Hindu, it is prana. To the Chinese, it is chi. In Japan, it is Ki. Mesmer called it animal magnetism. The author Bulwer Lytton called it Vril. Wilhelm Reich coined the term Orgone energy. In some ancient European traditions it is Pneuma. To Christian mystics, it is the Holy Spirit, or the breath of life. In the Druid tradition, it is Nwyfre (pronounced Nweevre) an old Welsh word meaning energy, vigour, vivacity or firmament and related to the proto-Celtic word Naomh - heaven. The French word for life, la vie, and the verb vivre (to live), clearly have the same derivation, as does the name Gwenevere which literally means 'white breath of life.'

The notion that the Gwenevere apparitions associated with this discovery are a personified manifestation of the planetary or universal life-force is extraordinary. To find it is supported by something as straightforward as the original meaning of her name is even more extraordinary and suggests that I am not the first to make this connection.

Whatever you call it, the Universal Life Force seems to be at the core of most esoteric teachings. Developing and utilising the personal life-force as Chi is the central focus of the "internal" martial arts such as Chi Kung. To this end, most martial arts systems have a spiritual dimension incorporating meditation and internal generation of the Chi energy. On a more physical level, the power directed to break bricks and achieve other extraordinary feats of strength is also Chi, focused and directed purposefully.

Chi may also be directed for healing and in acupuncture, the healthy flow of Chi in the body's meridians sustains health.

More esoterically, the Indian Tantric system and eastern mysticism generally, recognise seven spiritual energy centres in the human body which are considered to be the vortices through which our personal spiritual energies link to those of the world around us. They are known by the term chakra, a Sanskrit word simply meaning wheel.

In some illustrations, the chakric system is depicted with patterns of sacred geometry applicable to each centre. Their seven locations are said to correspond roughly to the major endocrine glands which play an important role in human growth and development and help regulate a variety of bodily functions.

To actually see these energy centres requires a good degree of clairvoyance. The chakra exist, not in the physical body, but in the field of energies which comprises our spiritual essence.

The invisibility of the chakras to the average person does not seem to have prevented mystics and psychics by the score from consistently seeing, describing or illustrating them. Traditionally, they are likened to lotus flowers or circular vortices of energy and light, but to add further evidence to the enigma of the life force manifesting as female figures, the chakra are also represented by seven goddesses known as Shakti, who are the embodiment of each energy centre. The seven chakric centres are as follows;

1: Muladhara, the base chakra, located near the genital region.
2: Svadhisthana, the sacral chakra, located a little below the navel and known in the martial arts as the Hara, where Ki or Chi is said to be stored.
3: Manipura, the solar plexus centre between the navel and the sternum.
4: Anahata, the heart centre.
5: Visuddha, the throat chakra.
6: Ajna, the brow chakra, also know as the third eye, relating to clairvoyance and visionary ability.
7: Sahasrara. the crown chakra, at the top of the head and relating to the pituitary.

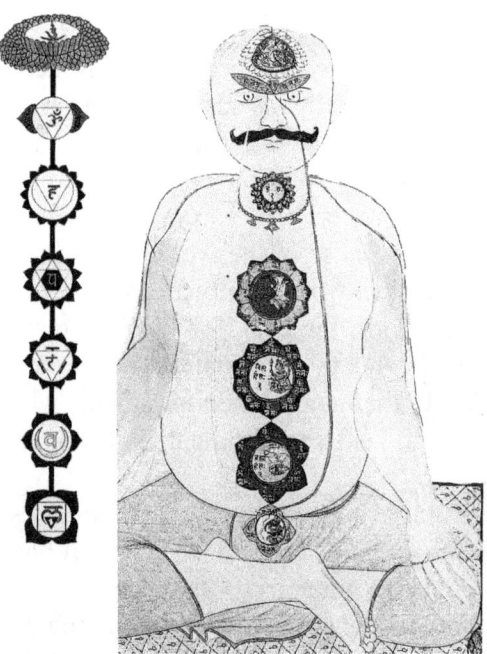

Illustration 117: Diagram of the seven chakra, the seven energy centres which link our personal spiritual energies (aura) to the spiritual dimensions of our environment.

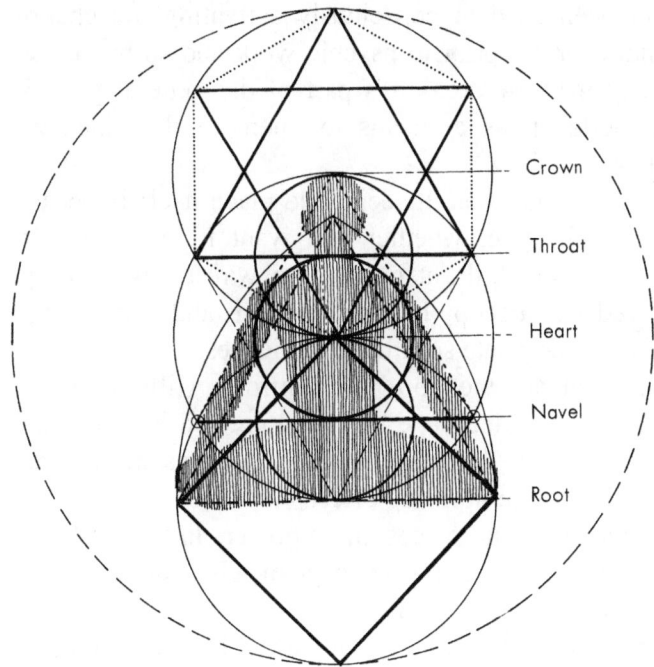

Illustration 118: The seven chakra may be depicted as a pattern of sacred geometry which places man as an intermediary between the spheres of the spiritual realm and the Earth.

As well as being fundamental to practices like Kundalini Yoga and Reiki healing, learning how to use the energy of these centres has become the basis of psychic and spiritual development systems commonly used by mediums, spiritualists and psychics.

These examples, it has to be said, do not simply relate to the spiritual energy of the human body In a broader context, they are all part of a holistic system. An understanding that the life force is universal and all-pervading is implicit. It is out there in the landscape, as well as inside our bodies and the two interact.

I have experienced this personally. One day, at Primrose Hill in north London, I felt a distinct energy tingle in the legs and base chakra as I approached the hill from the Queen's Pub gate at the foot of the hill. As I walked up to the summit, the sensations moved up my body, seemingly energising each of the chakric centres in turn, so that by the time I reached the crown of the hill, the energy had activated my crown centre. On other occasions I have followed the same route and felt little or nothing out of the ordinary, so there is a probably a cyclic and seasonal element associated with these effects. They are not constant and even when active remain undetected by the casual visitor.

I had been used to consciously activating the chakric centres as a meditational exercise prior to psychic work and so had developed a degree of sensitivity in those areas. It's part of the general psychic and spiritual training undertaken as a means to open up the channels for intuition, inspiration and healing.

At certain locations, it seems the place itself is occasionally capable of doing this for you, whether you want it to or not. If it's an effect exclusive to certain hills, it goes some way to explaining why so many natural sacred sites occupy hill top locations and why such places might be more likely to precipitate spiritual experiences.

Tuning in at the top of Primrose Hill, I realised that these prominences in our landscape had been created by huge upsurges of force and that a less physical element of that force was still active at these places and responsible for an uplifting effect.

In other times and places, these powers in the land were well-known. In the east, for instance, the concept of Chi extends to the landscape at large in the form of Feng Shui.

At the time of writing, Feng Shui is undergoing something of a renewal as an esoteric adjunct to interior decor, aimed at creating harmonious home and work environments which have a beneficial influence upon their inhabitants.

Its origins, though, actually lie 3,000 years earlier as a science which was used, not only for locating the most auspicious spots for temples, tombs, houses and other additions to the landscape, but to re-shape vast tracts of countryside to create an environment totally at harmony with the forces of nature.

In Feng Shui, Chi is said to flow along Lung Mei, or dragon paths. Within it are two opposing polarities, symbolised by the blue tiger and white dragon. The local Feng Shui practitioner, being a person who detects, deflects and generally works with dragon paths, is frequently known as a Dragon Man.

Like Feng Shui, sacred geometry acknowledges the unity of all things and implies a geomantic awareness of the spirit or energies of the land and the correct placement of sacred sites in relation to them. After all, it is a practice enabling mankind to plug into the energy of the land at its most vital power points.

The existence of Feng Shui confirms that a science based on an awareness and use of the Earth's spiritual energies did actually exist in the East (and still does). What is even more surprising is that similar systems were widely practised in other parts of the world, including Europe and the

British Isles. For this information I am indebted to Nigel Pennick, whose thoroughly researched book, EARTH HARMONY, documents several of these geomantic sciences.

Vastuvidya, for instance, is a Hindu system which governs the location and design of sacred places or buildings with the aim of harmonising spiritual and physical aspects of the environment.

Another system practised on Madgascar is known as Vintana. According to Nigel Pennick, it derives from traditions originating in Malaya, Islam, China and Africa but merges them into a composite system which is still used today to determine the most auspicious sites for homes as well as their internal layout.

The European system mentioned by Nigel Pennick, evolved from the eight directions (our modern compass points) and eight natural divisions of the year; The two solstices (the longest and shortest days), the two equinoxes (the days of equal light and dark) and the four quarter days which were celebrated as Celtic fire festivals (see the wheel of the year illustration in a earlier chapter).

It has to be said that, as far as I know, none of these systems incorporate any complex notion of precise landscape geometry on a grand scale, but they do incorporate practices from which the applications of sacred geometry in architecture may have evolved.

The Manasara Shilpa Shastra's method of laying out a temple groundplan involves a geometric division into four, then eight, related to the directions from the place of power, so this may prove to be a universal basis common to all systems.

Feng shui also makes use of the eight directions in the form of the Pa Kua although Feng Shui's dragon paths, as the name suggest, are meandering serpentine routes rather than linear alignments with geometric connections.

Feng Shui's only straight lines are man-made and are laid out on the landscape emanating from the Emperor's palace. They symbolise his "rule" over the land proving that straightness is associated with rulers in both senses of the word.

It is probable therefore that an attempt to contain, control and direct the energies of the land may be behind the idea of laying out an area with precise patterns of various types.

Of course, alignments of standing stones, megalithic monuments, leys and other associated phenomena elsewhere in the world may be evidence of similar practices, as John Michell and other ley researchers suggested long before these ideas seeped into my thoughts.

Some supportive evidence for this notion is to be found in Feng Shui where one of the basic methods of harnessing the dragon currents of the land is to "fix the dragon in one spot" by knocking a stake into the ground.

Now this has interesting correspondences elsewhere since a dragon or serpent is used world-wide as a symbol to represent the powers of the Earth. More importantly, it echoes the procedure followed in the establishment of the earliest temples described in the last chapter.

In Britain and Europe, we have plenty of sacred sites dedicated to dragon slaying saints which may also be founded on some association with this geomantic process. St. Michael's dragon slaying prowess may derive from acts which re-pinned the dragon, or Earth-current, at a site prior to its Christianisation.

Certainly, the European geomantic systems based upon the eight directions (sometimes also called eight winds or eight spirits), began with the establishment of a central point as either a pole, gnomon, tree or omphalos stone and conceived as a world-axis relative to the local landscape. The development of the eight directions from it would almost certainly have resulted in similar geometric patterns to those used in the establishment of the earliest Hindu temples.

This geometry has already been shown to be at the basis of the various practical applications of sacred geometry in religious architecture. It is probable therefore, that these practices were once widespread. It is worth mentioning again, in this context, that the concept of eight directions is fundamental to Feng Shui in the form of the Pa Kua and in the construction of the Lo-Pan, or geomantic compass traditionally used by dragon men.

The eight directions may also be fundamental to Stonehenge's forty stone ring. If its inherent geometry was conceived as a forty-point star based on a pentagram directed to each of the eight directions, it may have been conceived from this kind of geomantic system and its symbolism may relate to its use as a magical device to harness and control the spirits or forces of the eight winds.

Equally, with a pentagram controlling each of the four compass points, the Earthstars twenty point star may have similar functions. The fact that it is on the circle of the Earth Spirit makes this symbolism doubly appropriate.

The existence of some kind of subtle energy at sacred sites or flowing between them has kindled many heated debates over the years. Dowsers are able to detect lines of energy within the landscape which they invariably regard as spiritual in nature. Some, notably Colin Bloy, Clive Beadon and Michael Poynder have also detected numerous geometric patterns.

Unfortunately, I am not what you would call an experienced dowser, although I do have some ability in that direction. I can literally feel the energy in my hands at certain places, sometimes as a faint tingling, sometimes as a strong throbbing pulsation with a sense of movement and direction to it. On several occasions I have actually felt the force of an alignment flow through me as a double spiral of energy moving across the Earth's surface

Such evidence, however, is subjective and therefore inadmissible to the unconvinced. The sceptical are entitled to their opinions and I don't think it's worth arguing with them. I know what I have felt and no amount of describing it or trying to rationalise it will sway the opinions of someone who hasn't felt the same thing and is not overly inclined to believe me.

An important consideration is that it's a phenomenon with a spiritual basis. Science is not yet capable of building a device to detect the human soul or any other spiritual energies. It has not yet come up with an E.C.G. machine or stethoscope that can tell what someone feels in their heart.

We can sense these things empathically, because our essence is the same stuff. Of necessity, it has to be a subjective process.

Our sacred sites and holy places somehow provide locations where we can tap into the spiritual energies of the world around us, or be influenced by them consciously or subconsciously.

Whether they flow along laser-straight paths between sites, or spiral through sweeping arcs and circles between them, or both, is open to debate.

The life force is everywhere. Its specific trajectories may actually be numerous and ultimately unfathomable to the human mind.

What is more certain is that recognisable geometric patterns in the Earth's structure have to be the result of holistic patterning which transcends all considerations of scale and exists at every level from sub-atomic structures only visible via electron microscopes to universal designs inscribed for all to see, in the heavens above.

AS ABOVE, SO BELOW

To those of us who merely glance at the starry skies and see a confusion of sparkling lights, it may come as a surprise to learn that the orbits of certain planets when viewed from the Earth describe perfect geometric patterns in the heavens.

Confirmation of this notion comes from the work of the German astronomer Joachim Schultz. His book, MOVEMENT AND RHYTHMS OF THE STARS, sub-titled A GUIDE TO NAKED EYE OBSERVATION OF SUN, MOON AND PLANETS, reveals that Venus' movements, for instance, trace a perfect five-point star in our skies over a period of eight years. Note that five and eight are linked here again as they are in Stonehenge's geometry and in London's Earthstars.

Mercury cuts a hexagonal figure in the night sky over a twelve month period, while amongst Jupiter's patterns is one that traces a circle of thirty loops in the night skies.

So while sub-atomic particles whirl through paths of sacred geometry in their orbits around their nuclei, the stars and planets whirl around the universe in similar patterns.

As above, so below.

There may even be a more direct link between geometric star patterns in the Earth and the stars above. Just as the gravitational pull of the moon creates tidal flows in the Earth's oceans, other planetary bodies may have stirred their own subtle tides when the Earth was no more than a ball of molten white hot lava, its surface features as yet unformed.

These tides may still ebb and flow in the Earth's subtle energy fields as well as in its malleable mass. Who knows what part they may have played in creating the Earth's contours, setting the stage for our sacred sites? Who knows, too, what empathic influence they may still have. Those tidal imprints may still pull deep within the Earth. The centre and bulk of our planet is still molten. The crust we live upon is thinner than an eggshell, relatively speaking.

The path of the sun would have had a particularly strong influence, especially at the times of the midsummer or midwinter solstices when it would appear to rise and set at its most northerly or southerly positions over an extended period. This could account for the fact that mid-summer and mid-winter sunrises do seem to have played an important role in the establishment of ancient sacred monuments, the world over.

The axis of Stonehenge, as I am sure you know, is oriented towards the midsummer sunrise/midwinter sunset alignment close to the heelstone while the entrance passage of the huge and ancient mound at Newgrange in Ireland was constructed to enable a shaft of sunlight to enter it precisely at the moment of sunrise on the midwinter solstice.

THE HORSENDEN HILL SUNRISE LINE

The Earthstars geometry also embodies several major alignments which appear to match midsummer sunrise and sunset angles.

The outer line of the original pentagonal figure, from Horsenden Hill to St. Mary's East Barnet is a midsummer sunrise line. When viewed from Horsenden Hill, the sun rises along this alignment. St. Mary's East Barnet is not itself visible from Horsenden Hill. Barn Hill at Wembley obscures the view. The alignment is valid nonetheless. I have witnessed the midsummer sunrise here on many occasions and each time, felt a surge of solar force which I believe energises the entire pentagonal patterns annually through this and similar alignments.

Interestingly this insight goes some way towards validating the psychic message I picked up, calling Horsenden the Hill of Horus. In Egyptian mythology, Horus is also known as the Hawk of Dawn, representing the power of the sun at daybreak, the Hill of Horus is therefore an extremely appropriate name for a place which is energised by a sunrise alignment at midsummer.

Further along the alignment at St. Mary's East Barnet, the midsummer sun rises directly over an impressive white building on the opposite side of Oak Hill Park. It is actually another religious establishment, sited directly on the line; a Theological College with its entrance in Chase Side, Southgate.

At St. Mary's, the midsummer sunrise alignment can easily be found at any time of year. The largest and most impressive monument in the churchyard, a sort of gothic rocket spire over twenty feet high, is located beside it and acts as a permanent mark point for the alignment in much the same way as the Heel Stone at Stonehenge.

Across the road, in Oak Hill Park, it is not what is on the alignment that is significant so much as what is missing. There is a large gap in the grass verge which suggests that Oak Hill Park was once fenced and at the precise point where the midsummer sunrise line passed through it, was once a gate. The same phenomenon can be seen at Denham's St. Mary's church where a ley alignment actually passes through a gate in the church and the entrance gate to Denham Place. This is a common feature of ley alignments and was first noted by Alfred Watkins in his early books on the subject.

The pentagonal grid contains several alignments parallel to the Horsenden to St. Mary's line. These may also be simultaneously energised by the sunrise. In the main pentagram, the line from Caesar's Camp Wimbledon to St. Gabriels Wanstead, for instance.

Once this orientation is recognised, it is astonishing to see how many of London's streets have been built in close alignment to it.

In Paddington, there is a whole block of streets around Sussex gardens aligned to a midsummer sunrise. St. James' church seems to have been deliberately placed at the end of Sussex Gardens as the focus of the alignment. Several other sunset/sunrise alignments in London are worth mentioning.

From Westminster, on midsummer's day, the sun would appear to set in the area of Harrow-on-the-Hill. From St. Pauls, a similar alignment picks out the midsummer sunset point as Primrose Hill. In the opposite direction a path in Greenwich Park is closely aligned to St. Pauls and the midsummer sunset axis. Obviously, midsummer sunset lines double as midwinter sunrise alignments and vice versa, depending on your viewpoint along the line.

Such important celestial events would have identified these places as immensely important ritual locations for our ancestors.

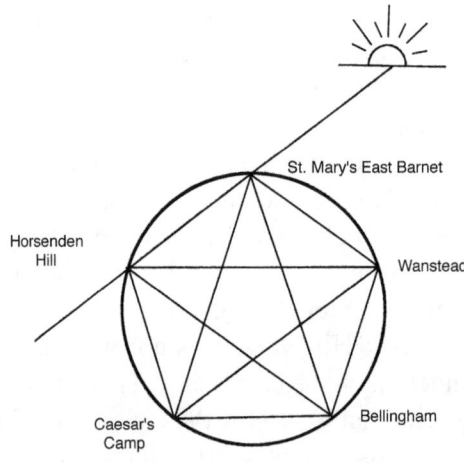

Illustration 119: The midsummer sunrise line between Horsenden Hill and St. Mary's, East Barnet, in relation to the pentagonal patterns of London's Earthstars.

Illustration 120: The midsummer sunrise at St. Mary's East Barnet. The shadow of the author indicates the sunrise line and highlights some curiosities. First, the most prominent monument in the churchyard which is on the alignment and acts as a sighting-line marker, rather like the Heel Stone at Stonehenge. Secondly, a gap in the grass verge opposite the church indicates that a gate into Oak Hill Park once stood directly on the midsummer sunrise alignment.

THE CORONATION LINE.

An even more impressive alignment which merits a separate mention, is a possible midsummer sunrise alignment through Westminster Abbey. It identifies Ludgate Hill as its sunrise point.

So from the region of Westminster Abbey on the longest day of the year, the sun would appear to rise out of Ludgate Hill beneath where St. Paul's now stands. Conversely, from St. Paul's at the midwinter Solstice, the shortest day of the year, the sun would appear to set in the area of Westminster Abbey.

This would have been of enormous significance to our ancestors to whom the rising and setting points of the sun and moon may have been akin to the birthplace and resting place of their gods.

It may even explain why their relative locations were chosen as sacred sites at some time in our distant past.

A little further along the same alignment are some other notable sites: St. Vedast's Church in Foster Lane; the remains of the extremely ancient St. Mary's Aldermanbury at the corner of Love Lane and Aldermanbury, which still has a highly charged atmosphere and is an very pleasant spot; The Chapel of The Open Book in Sun Street. Last but definitely not least, is a mound of uncertain antiquity at Shoreditch, in the centre of Arnold Circus. Unlike many of London's ancient monuments this one didn't disappear beneath urban development. It has been thoughtfully incorporated into the traffic island at the centre of the junction. Several well-known ley researchers mention Arnold Circus in connection with one of Watkins' original Leys. However, they refer to it as the being "the former site of The Mount" so obviously never paid it a visit. The Mount is reduced in size, but still there, alive and buzzing with energy

One final consideration, the section of Holloway Rd (Holy Way Road?) which runs parallel to one of the pentagonal star lines is orientated directly towards this mound, so it could be an ancient ritual site of considerable significance.

In the opposite direction, this Westminster Abbey - St. Paul's alignment stretches to Kingston-upon-Thames which takes its name from the King Stone, the Coronation Stone upon which seven Saxon Kings were crowned. Thus today's coronation site of Westminster is linked to yesterdays' by a royal sunrise alignment. I suspect this may not be coincidental. In addition, the King Stone, as you may recall from a previous chapter, is an important mark point on the 30-point Earthstar.

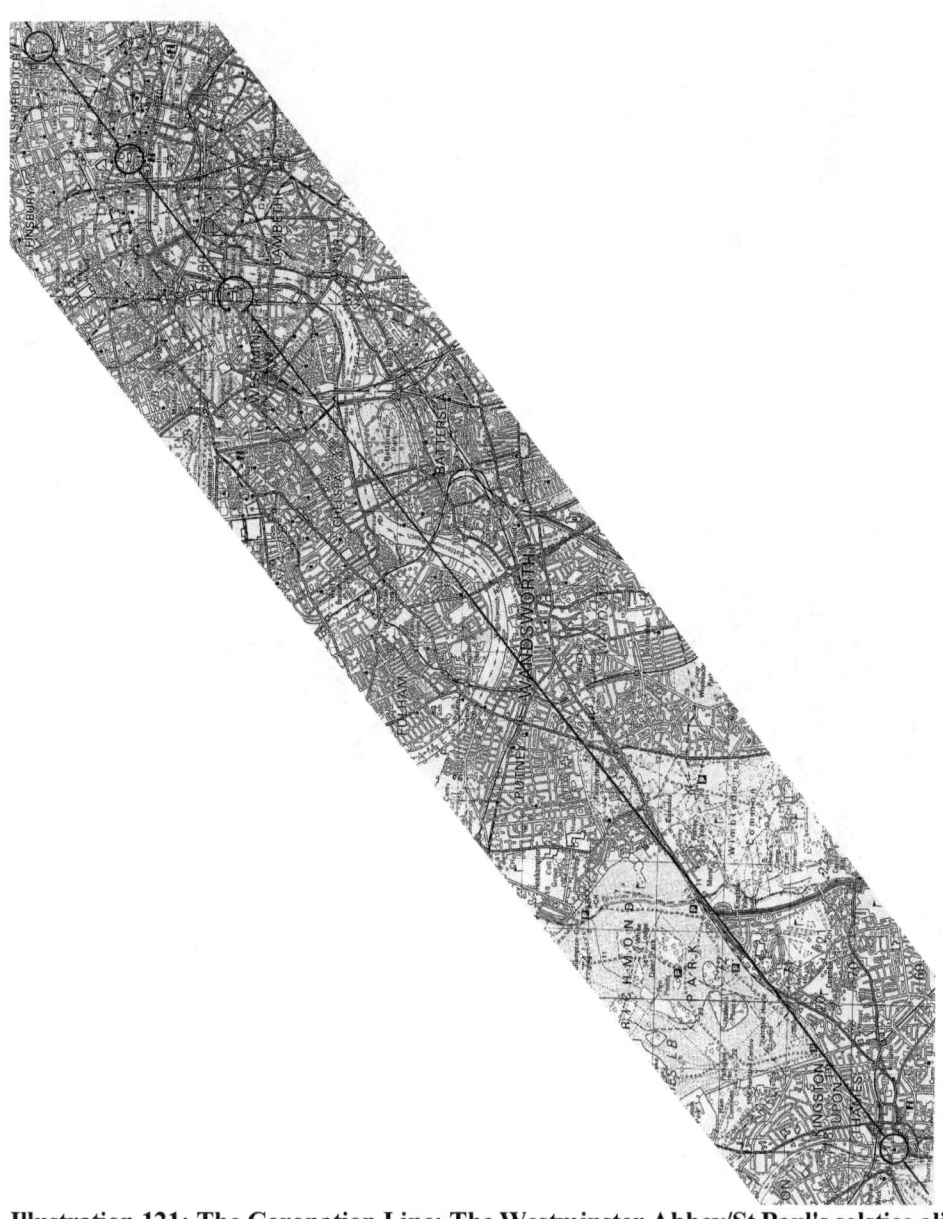

Illustration 121: The Coronation Line: The Westminster Abbey/St Paul's solstice alignment extends to the King Stone at Kingston-upon Thames, thus linking the Coronation Stone of the Saxon Kings with the current coronation site of Westminster Abbey. O.S. map © Crown Copyright. All rights reserved. License No. 100029558.

Illustration 122: The Mount, an almost forgotten ancient site, now incorporated into a traffic island at Arnold Circus, Shoreditch.

Illustration 123: The King Stone at Kingston-upon-Thames. The Coronation Stone upon which seven Saxon Kings of England were crowned. It is on the Westminster Abbey-St.Paul's Coronation Line and is a mark point of the 30-point Earthstar.

London   City of Revelation

Illustration 124: The Barking Abbey/Westminster Abbey line, marked by Barking Abbey, the Tower of London, Southwark Cathedral, Westminster Abbey, St. Mary the Boltons and several other sites. O.S. map © Crown Copyright. All rights reserved. License No. 100029558.

LONDON'S STONEHENGE LINE

This may be London's most powerful alignment. Its sites include Barking Abbey, the Tower of London, Southwark Cathedral and Westminster Abbey. These are four of the most ancient sacred sites in the capital. To find they are all linked in this way is remarkable evidence that alignments of this sort may be deliberately planned.

Evidence of its connections to the Earthstars design is readily available too. At Barking, the parallel road phenomenon re-appears. The line runs directly in line with the A124 Barking Road. In the vicinity of the Tower of London, the line passes directly through the Earthstars' pentagonal junction point on Tower Hill as well as the Tower of London itself which stands on the site of an ancient sacred mound known as the Bryn Gwyn, the White Hill, beneath which Bran's head was buried to act as an inner-world guardian of his realm, The British Isles.

Extending the line westwards confirms its importance. It can be tracked directly to Stonehenge, through a number of lesser places including Staines, whose name may derive from the stones of another ancient megalithic monument. The Stonehenge connection raises the possibility that the alignment originated from Stonehenge and is of similar antiquity.

Illustration 125; London's Stonehenge line. The alignment from Barking Abbey - through the Tower of London, Southwark Cathedral and Westminster Abbey can be tracked directly to Stonehenge.

LIFE'S RICH PATTERN

Obviously, there is a complex matrix of energies associated with the structure of the planet, some of which scientists are already aware, some of which they may be beyond their capability to detect with scientific apparatus. If the Earthstars geometry represents some part of it, a measure of its complexity may be judged by the thirty-pointed star. When all the possible chords between the points are made, a net immensely rich in strands results. Yet this is only one level of the patterns.

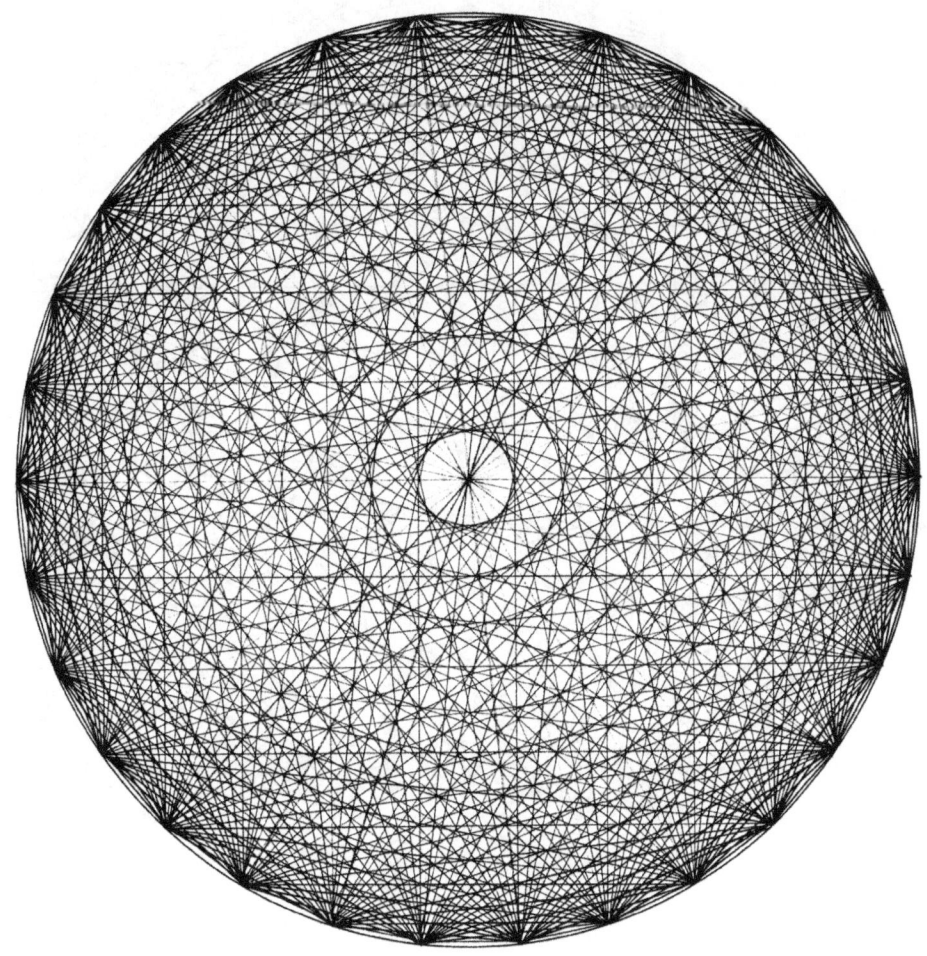

Illustration 126:

The complexity of the 30-point star with all possible chords connected. Even here it looks three dimensional and begins to suggest it may be part of a vast global grid.

The twenty-pointed star produces a net of almost equal density, as does the eighteen-sided figure.

All this is before the six, eight or eighteen-point figures have even been considered as overlays. Add the inescapable conclusion that whatever the complete design, it is actually three dimensional rather than just two and we can envisage a web of life that is all-pervading in its complexity, but comprehensible to us mere mortals, only in fragments.

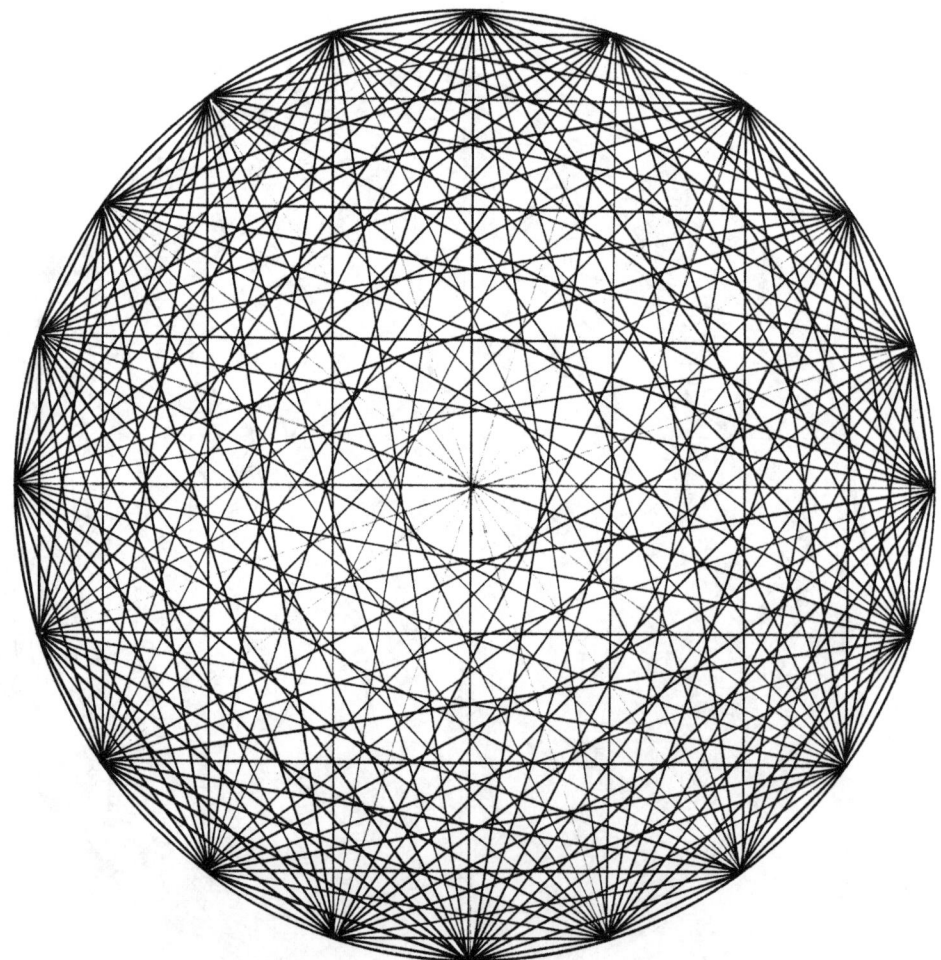

Illustration 127:
The complexity of the 20-point star with all possible chords connected. It could be another layer of the global matrix.

If our sacred sites and other key locations are built upon such a net, it is hardly surprising that some are found to form simple, straight alignments or other geometric figures. The average straight line ley would be, and is, a relatively common phenomenon.

Of course, when dealing with a network as complex as the Earthstars' grid, not all of the nodal points will have been built upon. Nor will they all fall upon noticeable natural features or places that present the correct physical attributes to create a special atmosphere. But if enough sites are identified, patterns may begin to emerge and be recognised.

Such designs do not necessarily have to have been planned for any mysterious esoteric purpose although I wouldn't rule out this possibility. The Earthstars patterns, as I have demonstrated, were known to the builders of both the Great Pyramid and Stonehenge and they both seem to have been constructed as an expression of a long forgotten ancient science.

On the other hand, the basic geometric forms are prolific within nature so they could actually crop up almost anywhere. Having said that, it does seem unlikely for them to combine in such spectacular complexity at random. Only in very special locations would such patterns emerge.

A good example of this is at Stonehenge where the four Station Stones of the monument are positioned to mark important solar alignments, the midsummer and midwinter sunrises and sunsets.

In addition, the Station Stones also form a perfect Golden Mean rectangle; a rectangle which could be formed by four points of a hexagon. Only at Stonehenge's precise latitude would those alignments create this exact geometric figure. A few miles further north, or south, and the same sighting lines would mark an irregular quadrilateral.

The same principles apply to London's Earthstar patterns. A design of this complexity and precision could not occur in a random location. It is hardly likely to be commonplace.

It identifies a very special place.

A vast temple of the stars set in a sacred landscape.

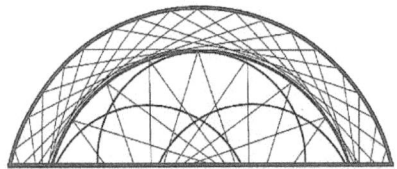

Chapter Eleven
Journey to the Centre

All geometric constructions start from a single point and in sacred geometry, that point, like a seed, is said to embody both the beginning and the final construction. This is certainly true of the Earthstars' geometry despite the fact that the centre was not one of my starting points. It was actually identified after most of the star patterns had been found.

When I first investigated the sites along the North-South axis between St. Mary's East Barnet and Pollard's Hill, I didn't realise the line's mid-point was marked by anything particularly significant.

The nearest church to the centre is Nash's classical All Souls at Langham Place, adjacent to the BBC's offices, although it is not actually on the alignment (It's on the 8-point diagonal). The closest religious establishment directly on the line is the Central Synagogue in Hallam Street. The centre of the Earthstars patterns falls a little further south, near to the junction of Hallam Street, Gildea Street and Langham Street.

One thing I did notice over the course of many visits was that this area may have been a much more prominent spot way back when it was open countryside. There's a noticeable incline downhill towards Marble Arch on the western side, south towards Piccadilly and, to a lesser extent, northwards towards Marylebone Road. To the east, in the direction of Tottenham Court Rd, it looks reasonably level.

I hesitate to describe this as a hill. It doesn't thrust itself intrusively into the skyline like the upturned pudding bowl of Silbury (which is directly on the Earthstars' main East-West axis), but in the days when office blocks didn't obscure the view, it may have provided impressive sighting lines to the horizon in all directions.

I intended to check old maps of the area to see if anything significant used to occupy the spot, but before I could, a letter from Mr. L. T. Street in Ruislip enlightened me.

He'd read my first Earthstars book and thought I'd be interested to know that a church dedicated to St. Paul had once stood within a few yards of the exact geometric centre of the Earthstars design.

It had been demolished in 1905 and an office block now stood on the site. He had actually worked in that office for over twenty years and for a great deal of that time, he informed me, the staff were not allowed to work on Sundays as it was still consecrated ground. The building is Brock House, currently used by the BBC. It stands detached from any other buildings on the original site of St. Paul's Portland Chapel, a rectangular plot between Gildea Street and Langham Street.

A trip to the Local History section of Marylebone Library and enquiries to the Corporation of London provided me with further information. The St. Paul's Chapel was not the earliest construction in the area. In 1668, a reservoir known as The Marylebone Basin had been built at this location to take advantage of the area's plentiful springs and provide water for the Covent Garden area.

In pre-Christian cultures, the life-giving waters of springs were often venerated and marked by a shrine. Many such spots evolved into today's sacred sites. Others simply disappeared beneath streets, offices, shops, houses or factories. Sadly these particular springs seem to fall into the latter category. No records survive relating specifically to them, so we know little except that they produced abundant supplies of water. That in itself is enough to suggest it would have been a well-known and well-respected spot to our forbears. It is worth mentioning, too, that many dowsers, in particular Guy Underwood, have noted the important involvement of underground water in the location of sacred sites and the unusual energies associated with them.

The Marylebone Basin reservoir lasted almost a hundred years. Presumably the springs were capped and diverted when the St. Paul's Chapel, designed by the architect S. Leadbetter for the Portland Estates, was built between 1760 and 1766.

An oddity worth noting is that the altar of the St. Paul's Portland Chapel was at the western end of the building instead of the more usual eastern end. No-one seems to know why, but as the Earthstars centre point is near the western end of the building, not the east, it may have some bearing on the matter. It reminded me that Guy Underwood's dowsing had discovered that church altars had frequently been positioned over blind springs or the crossing point of underground streams. What fascinated me more was the building's proportions.

I could find no ground plans, but discovered it had been 39.6 ft in width by 68 ft long. This is surprisingly close to the proportions of the original St Mary's Chapel at Glastonbury, which I knew from the works of John Michell, had been built to a specific geometric ground plan,

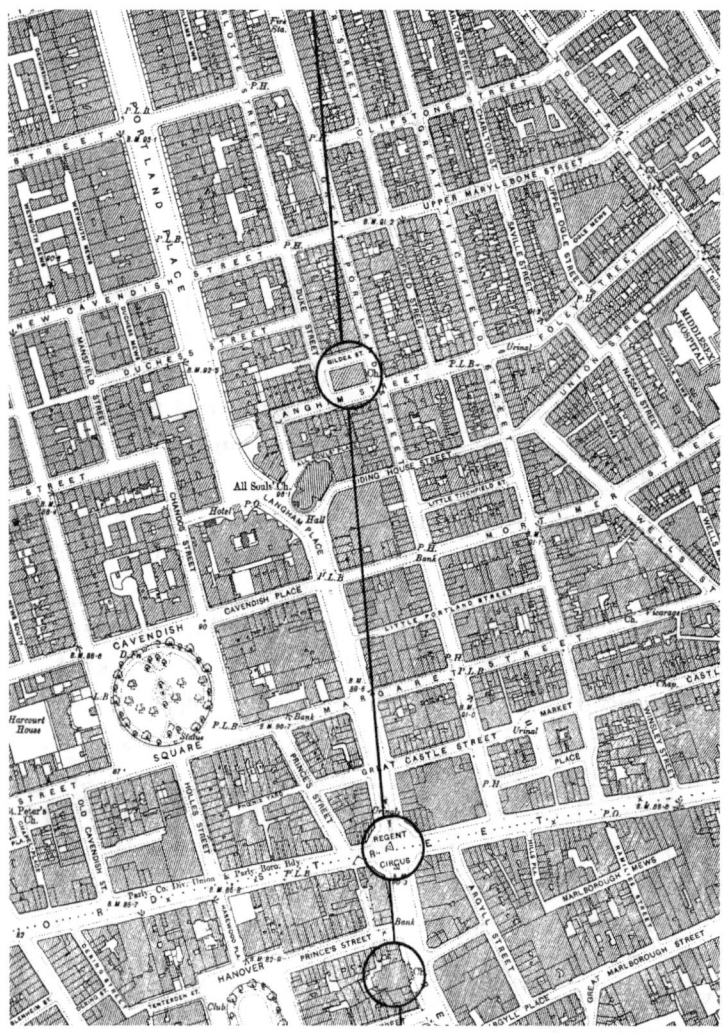

Illustration 128: Map showing the original St. Paul's Portland Chapel which once stood at the centre of the Earthstars' landscape temple.

Illustration 129: Engraving of the St. Paul's Portland Chapel which once stood at the Earthstars' centre.

Illustration 130: The present building on the site, Brock House, which is currently used by the BBC.

a golden mean rectangle whose four corners mark four of the six points of a hexagram (like the station stones of Stonehenge), or key points in an eight-point star plan.

The proportions of the St. Paul's Portland Chapel were identical in width and a mere six inches less in length. Close enough to suggest both could have been based on the same foundational design, taking into consideration the unknown width of its walls.

So what we have here is the remarkable discovery that at the centre of the Earthstars patterns was a church based on a design which embodied the very same principles of sacred geometry.

The chapel's intrinsic geometry mirrors part of the Earthstars design and being a construction exhibiting golden mean proportions, it would have resonated harmoniously with them.

This compatibility is obvious from the relative proportions of the figures which exhibit a harmonious relationship. The 39.6 ft width of the building fits precisely into the 16.2 miles diameter of the Earthstars inner circle 2160 times. 2160, if you recall, is the lunar diameter figure in miles and related to the solar mean diameter number 864 through a relationship with 4 (216 x 4 = 8640. 39.6 reflects the mean Earth radius number, 3960.

This is not to say the designer of either the St. Paul's Chapel, or the Lady Chapel at Glastonbury were aware of the Earthstars patterns on the landscape. An awareness of the principles of sacred geometry would have been enough to account for the correspondences since they would simply have been using traditional proportions compatible with the creation of the natural world and the universe around us.

This was, originally, an important consideration within the Masonic Guilds, whose trade secrets were perhaps not entirely derived from practical knowledge of applied building techniques.

Working from templates of sacred geometry ensured that man's additions to the landscape followed the same principles as the creators'.

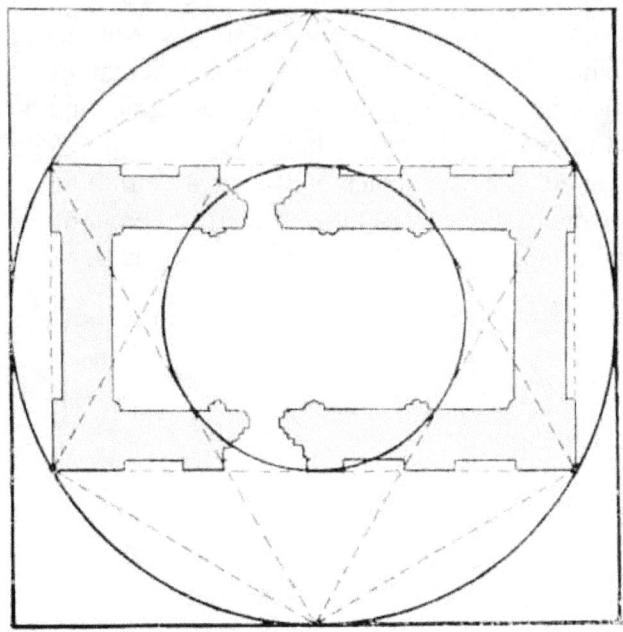

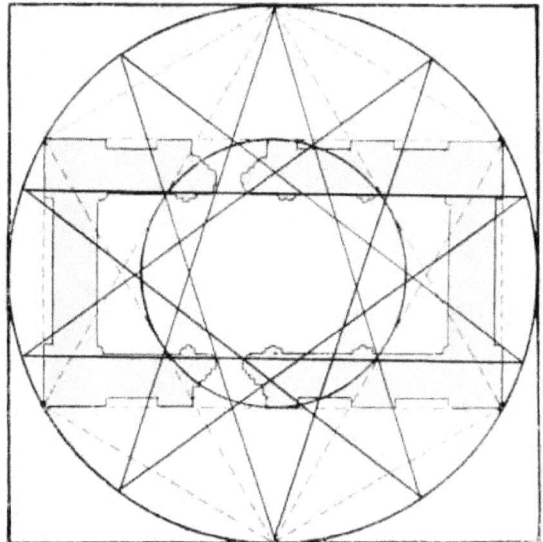

Illustrations 131 and 132: The Lady Chapel at Glastonbury Abbey is almost identical in proportion to the St. Paul's Chapel which once stood at the centre of the Earthstars Mandala. It shows signs of having been based upon a combination of sacred geometry master plans. The hexagonal defines outer form while the pentagonal proportions define inner limits.

MACROCOSM AND MICROCOSM

The designs of sacred geometry were believed to bridge the chasm between Heaven and Earth, to link microcosm and macrocosm.

A graphic example of this principle is inherent in many of sacred geometry's most important figures, like the pentagram and hexagram. Each define within themselves a smaller, but identical figure. Within the smaller pattern is yet another, even smaller, and so on, ad infinitum.

Little wonder then that these patterns were considered to exist throughout all creation, from the smallest scale to the largest.

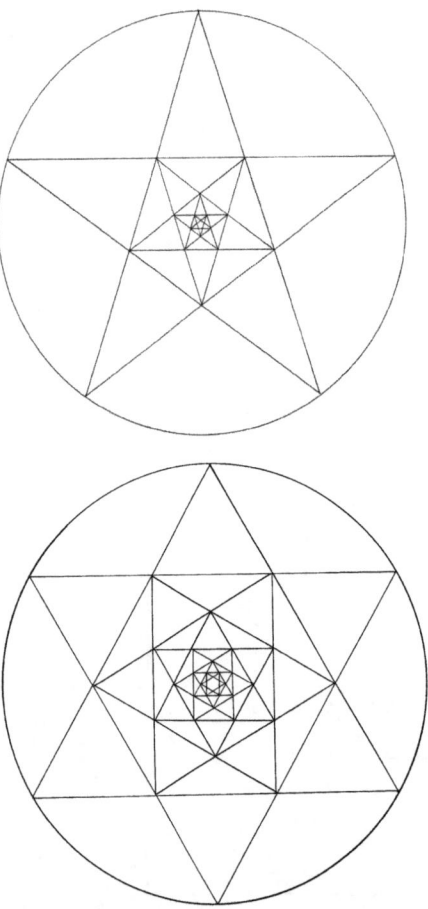

Illustrations 133 and 134: Every pentagram and hexagram contains an infinite number of identical figures within their geometry and so were thought to embody a link between macrocosm and microcosm

Of course, this applies equally to the Earthstars geometry. Within each of the Earthstars patterns are a series of identical designs, each progressively smaller. On closer examination they reveal some hitherto unsuspected geometric relationships between central London's principal sacred sites and places of historic interest.

The initial five-point star, for instance, contains an inner pentagon marked by St. George's in Patmore Street, Nine Elms; St. Andrew's at Frognal in Hampstead; an unmarked point in Abbottsbury Road Holland Park; Tower Hill over-looking the Tower of London; and an unmarked point on Highbury Hill overlooking a spot many Londoners consider sacred ground, Arsenal Football Club.

Among the sites identified by the five-point star within this figure is Westminster Abbey which lies on the line from Highbury Hill to St. George's Church at Nine Elms. The same line, incidentally, extends to the hill in Finsbury Park.

The Tower of London is on the line from Tower Hill to St. George's; St. Dunstan's and St. Bride's churches in Fleet Street, along with the London Stone, sit on the line from the Earthstars' centre to Tower Hill (this is part of the Horsenden Hill axis).

The circle of trees in Green Park figures prominently in this pattern, too. It helps prove its location is no accident by occupying the precise junction of two lines; The main N-S axis and the inner pentagram's line from Tower Hill to Holland Park which passes through the grounds of St. Mary Abbott's at the corner of Kensington High Street.

All this adds up to an impressive collection of important locations. Of course, there are also many places of lesser note to be found along these lines for those who care to examine the maps in more detail.

On an even smaller pentagram, we find St. Martin-in-the Fields, St. George's near Hanover Square, All Souls Langham Place and the British Museum, to name just a few.

The more complex patterns within the 10, 20 and 30-point stars identify Southwark Cathedral, Parliament Hill, Primrose Hill, Alexandra Palace, Crystal Palace's hill, the Hilltop in Brockwell Park and many other significant places; rather too many to mention, in fact.

Suffice to say that most, if not all, of London's ancient sacred sites and holy hills are to be found upon these alignments.

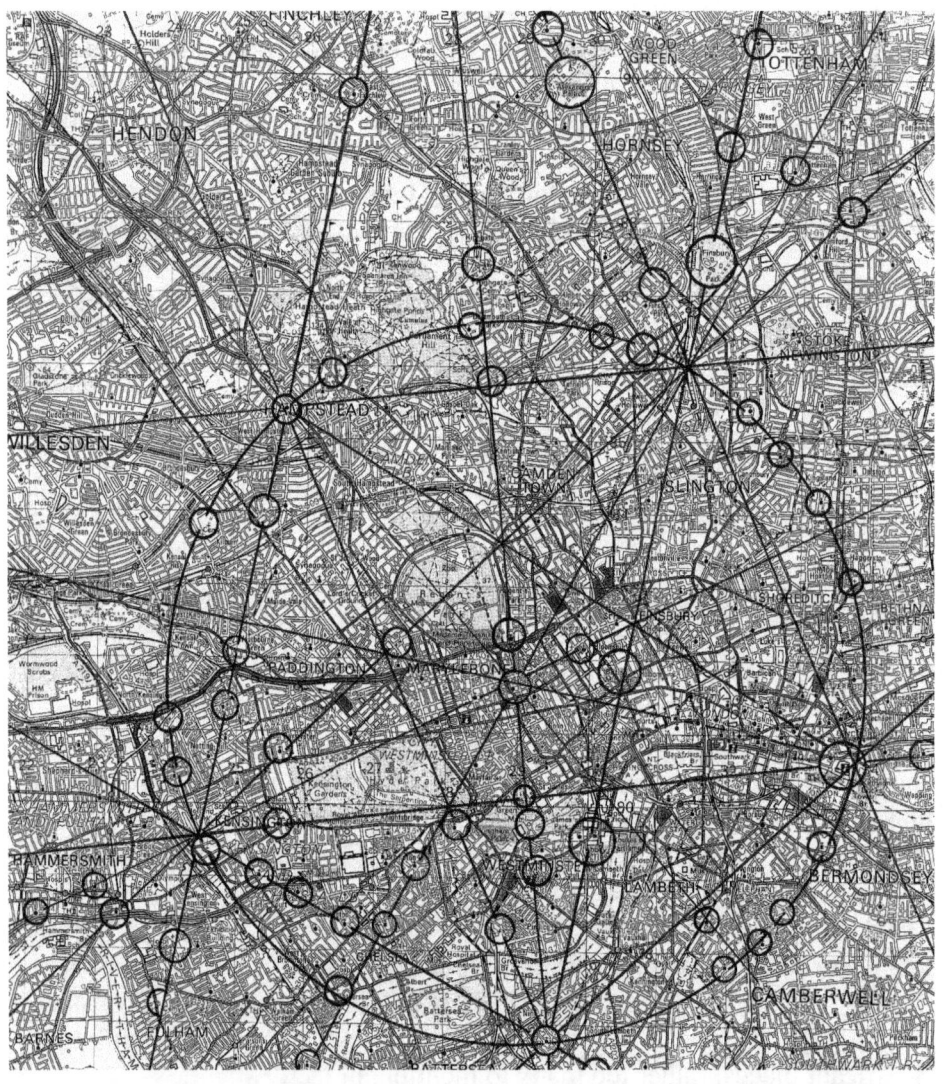

Illustration 135: Map of the inner pentagram defined by the intersections within London's original five-point Earthstar. O.S. map © Crown Copyright. All rights reserved. License No. 100029558.

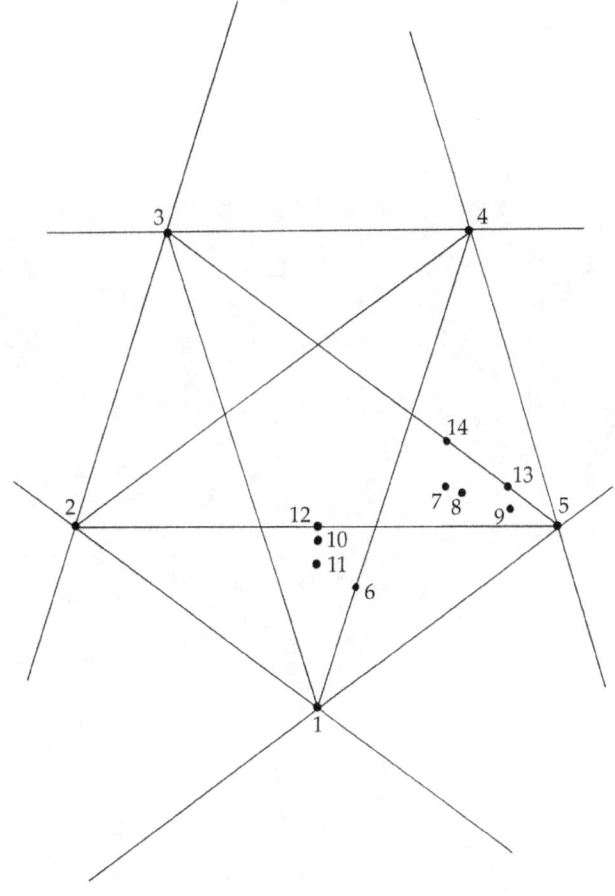

Illustration 136: Diagram showing the inner pentagram's main mark points.

    1: St. George's, Patmore Street, Nine Elms,
    2: Abbotsbury Rd near Holland Park.
    3: St. Andrew's Church, Frognal, NW3.
    4: Highbury Hill.
    5: Tower Hill.
    6: Westminster Abbey.
    7: St. Dunstan's, Fleet Street.
    8: St. Bride's, Fleet Street.
    9: The original site of the London Stone.
    10: The Victoria Memorial Fountain.
    11: Westminster Cathedral.
    12: The Green Park Grove tree circle.
    13: The Guild Hall.
    14: Church of St. Bartholomew The Great.

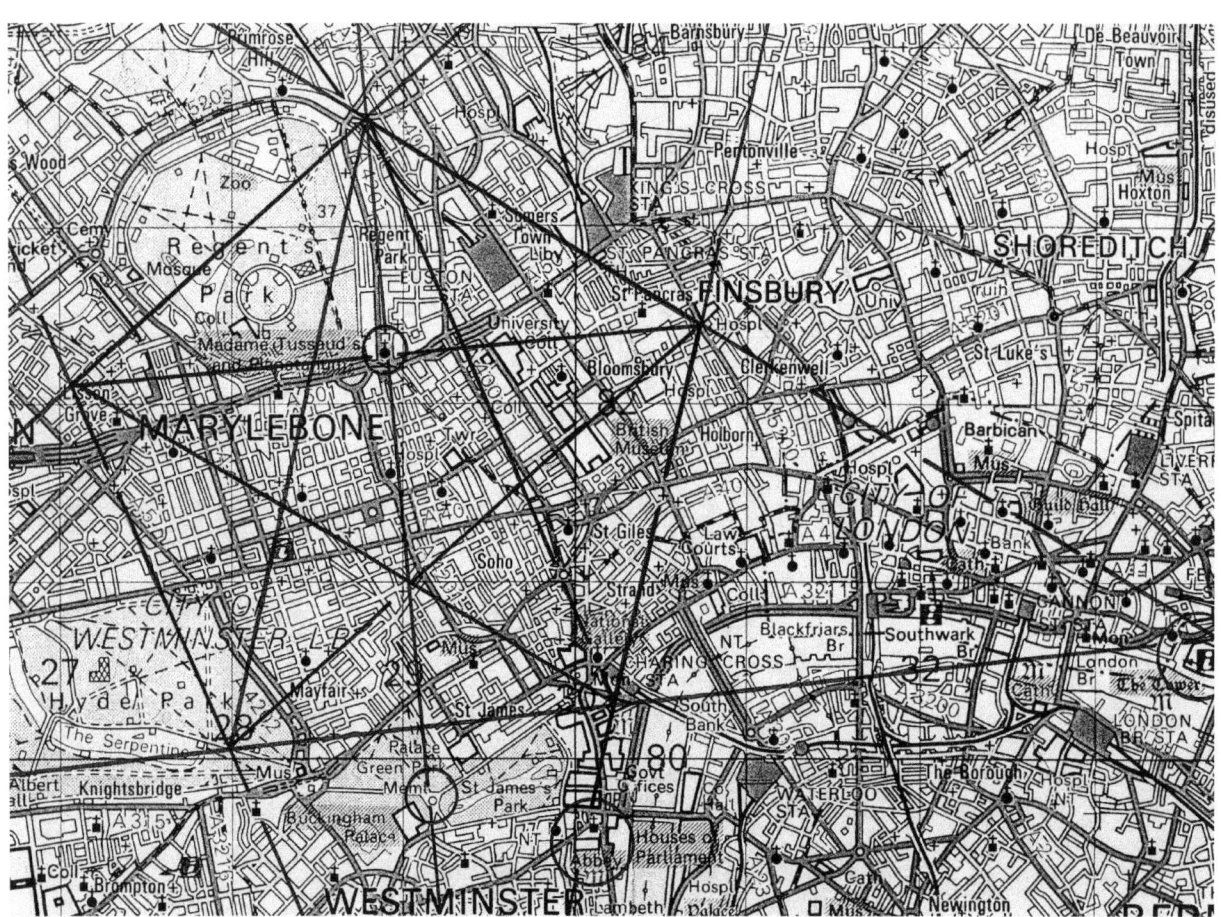

O.S. map © Crown Copyright. All rights reserved. License No. 100029558.

Without doubt, the most noteworthy pattern which can be replicated at a smaller scale is the foundational diagram of the Earthstars' two concentric circles.

This is important for two reasons. The first, you already know: As the squared circle construction, it is the basic framework upon which all the other Earthstars patterns are built.

The second will come as a great surprise. It certainly did to me. The proportions of the two smaller, concentric circles stem directly from the locations of the capital's two most well-known sacred sites, Westminster Abbey and St. Paul's Cathedral.

Taking the Earthstars centre as a starting point, the two circles can be constructed by simply taking a radius to Westminster Abbey, then another to St. Paul's.

The distance to Westminster Abbey is approximately 1.5 miles and the circle's circumference cuts through the centre of the main building diagonally, SW to NE. The distance to St. Paul's is 1.908 miles assuming the circle passes through Wren's dome.

These two circles are in the same ratio of 1:1.27 as the two original Earthstars circles. So the foundation of the Earthstars design occurs on the London landscape not once, but twice. In both instances, focused around the same geometric centre.

Not only are the two patterns identical in their proportional relationships, they embody a harmonic relationship to each other in their scale. The Westminster Abbey/St. Paul's circles would fit into the original Earthstars circles exactly 10.8 times. 108 is the lunar radius number. It is worth mentioning at this point that the Westminster Abbey circle's 1.5 mile radius is exactly 7920 feet. The same number in feet as miles in the Earth's diameter, so a harmonic relationship with the Earth's dimensions is also implied.

To have found this unusual and important geometric configuration once on the London landscape is astonishing enough.

To find it again, defined by the relative location of Westminster Abbey, St. Paul's Cathedral and the Earthstars' centre seems to confirm

Illustration 137 (opposite): Map showing the even smaller 5-point star within illustration 135's star. (O.S. © Crown Copyright). Some of the main mark points are as follows; Christ Church, Brick Street, W1; Holy Trinity Church opposite Great Portland Street tube station. The American Church, Tottenham Court Rd; St. George's Hanover Square; Green Park tree circle; St. Mary's Church, York Street.

beyond a shadow of a doubt that this is no chance occurrence. It is hard evidence and its implications for the reality of the Earthstars network are immense.

The Earthstars centre may not currently be marked by anything more significant than a hypothetical point on an ordnance survey map, but its existence cannot be denied.

Without it, this geometric relationship between St. Paul's and Westminster Abbey may never have come to light.

They are clearly interdependent parts of the same grand design.

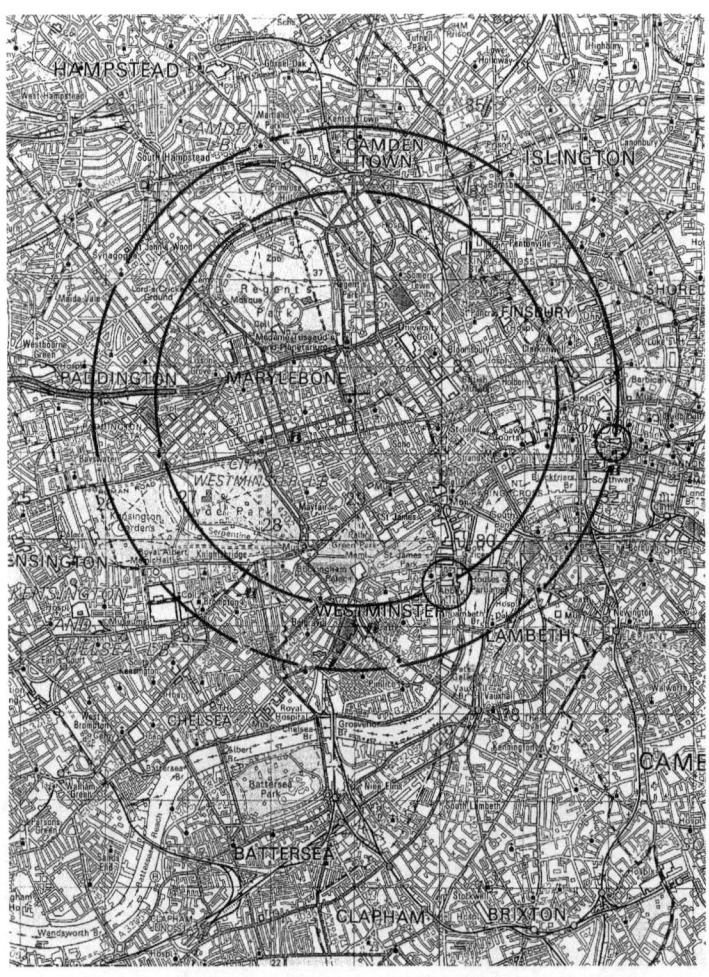

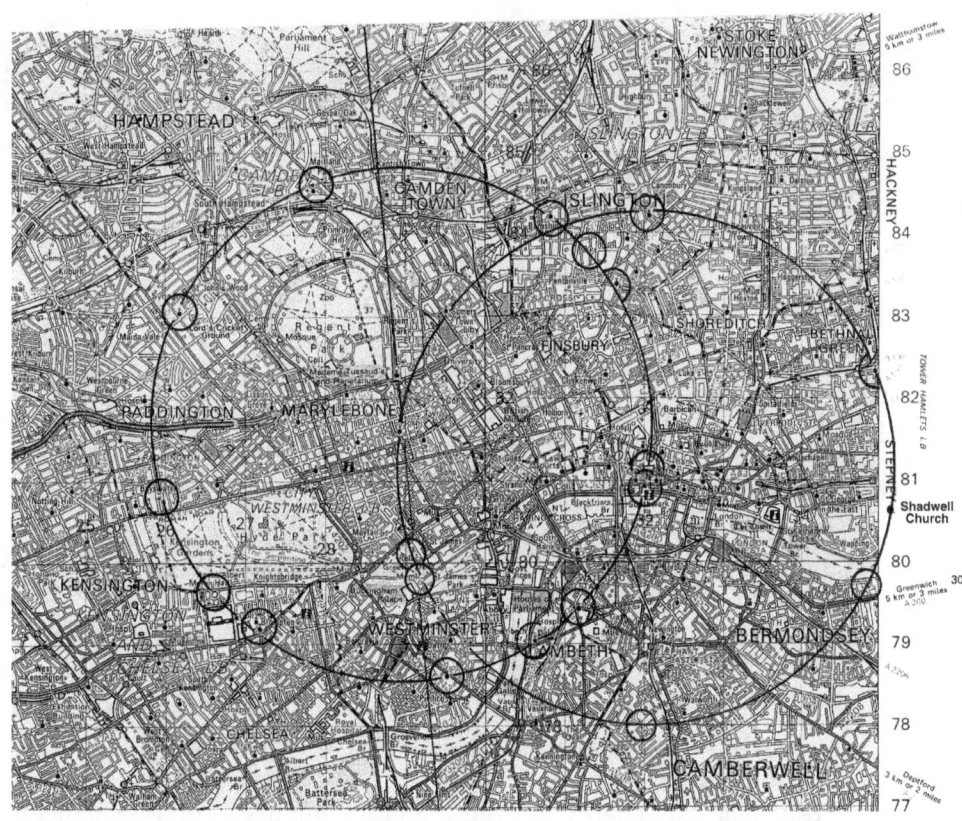

Illustration 138 (opposite): The relative locations of Westminster Cathedral, St. Pauls' Cathedral and the Earthstars' centre define a squared circle foundation identical in proportion to the Earthstars' patterns. It forms an intrinsic part of the complete Earthstars' design.

Illustration 139 (above): The relative locations of the Earthstars' centre point and St. Paul's Cathedral form an interesting vesica piscis which identifies many significant mark points. O.S. map © Crown Copyright. All rights reserved. License No. 100029558.

A more complex pattern between these sites can be found by centring a compass on St. Paul's and creating circles with a radius to the Westminster Abbey and the Earthstars' centre. The vesicas thus formed finds Westminster Abbey at the southern-most junction point of one and Lambeth Palace at the junction point of the other.

Many other prominent places spring to significance on these new patterns too. The ancient and atmospheric St. Etheldreda's at Ely Place, for instance, falls on the same circle as Westminster Abbey, as does the round Temple Church off Fleet Street and St. James's at the end of Sussex Gardens. On the Westminster Abbey and St. Paul's circles, we find St. Benet's Church at the corner of White Lion Hill and Queen Victoria St; Christ Church, Westminster Bridge Rd, Lambeth Palace, St. James the Less Church in Moreton Street near Vauxhall Bridge Rd, the Victoria and Albert Museum, the Royal Albert Hall, the Round Pond in Kensington Gardens, Our Lady Queen of Heaven Church off Queens way in Bayswater, St John's Church in Duncan Terrace, Islington, the site of the Kings Cross (at King's Cross), St. Mary-le-Bone Church and St Botolph's Church in Aldersgate.

A natural progression of this geometry reveals a common geometrical relationship between a great many of central London's major sites and suggests that the energy of these places might involve a concept of interacting spheres of influence and geometric relationships as well as simple linear ley alignments. One particularly interesting example is a perfect equilateral triangle formed by the Green Park tree circle, St. Martin-in-the fields and Westminster Abbey. In Illustration 142, the sides of the triangle have been extended to show that they form part of other alignments. One from St. Bride's in Fleet Street, through St. Martins, the Green Park Grove, the Wellington Arch at Hyde Park Corner, St. Paul's, Wilton Place and Holy Trinity Church Brompton. The Westminster Abbey to St. Martin's side of the triangle extends to Hawksmoor's St. George's in Holborn. A further interesting alignment which can be seen in this map is Regent Street, St. James' church in Piccadilly and Duke of York Street, all aligned to the triangle's junction point, the twin towers of Westminster Abbey's west entrance.

Central London's overall pattern is extremely complex and multi-layered. To explain it clearly in detail would take an entire volume, not just a couple of paragraphs. Rest assured, much more will be revealed in another book already in development. In the meantime, the patterns already disclosed (and their implications) are more than enough to keep us occupied for the time being.

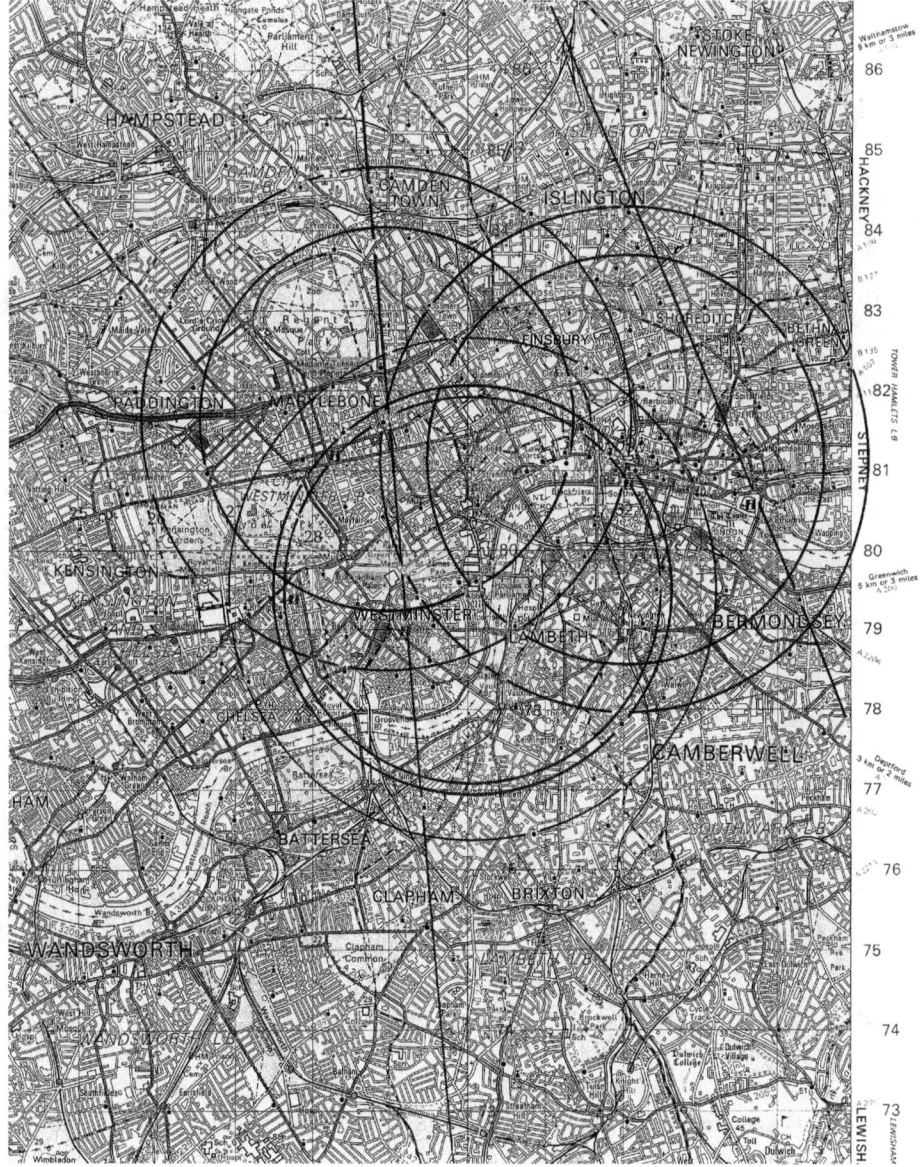

Illustration 140: Circles centred on Westminster Abbey and St. Paul's, radius to each other and the Earthstars centre creates this interesting triple vesica which counts Lambeth Palace, the Temple Church, the Victoria Memorial Fountain, St. Martin's on Ludgate Hill and St. Pancras Church amongst its mark points. Westminster Abbey sits at the junction point of one vesica, Lambeth Palace at the junction point of another. O.S. map © Crown copyright.

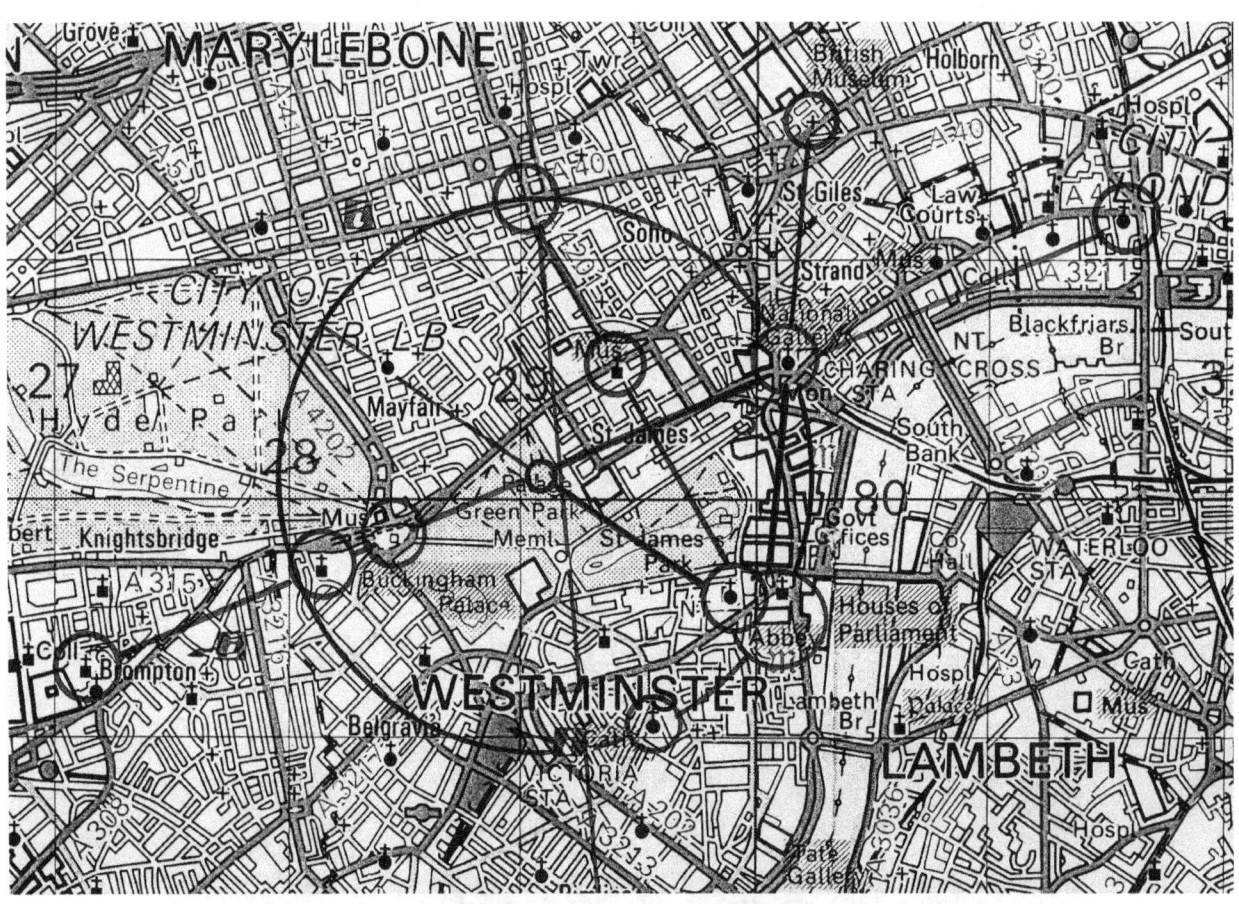

Illustration 141: An equilateral triangle created by Westminster Abbey, St. Martin-in-the-Fields and the Green Park tree circle. Note also that Regent Street, St. James Church Piccadilly, Duke of York Street and St. James Square are all aligned to Westminster Abbey through the Methodist Central Hall. O.S. map © Crown copyright.

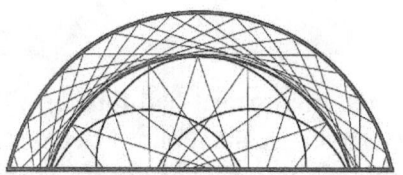

Chapter Twelve
Other Star Systems

Are there any precedents for the Earthstars' landscape geometry? Back in the early 1980s, when the discovery first came to light, I was aware of only one.

In a publication entitled GLASTONBURY, A STUDY IN PATTERNS published by RILKO (The Research Into Lost Knowledge Organisation) was an article written by John Michell. It was illustrated with a full page map of the area around Glastonbury Abbey over which had been drawn two circles interconnecting in a vesica piscis. Inside each circle were hexagonal star patterns evolving from the vesica's intersections.

As far as I remember, this was the only example of landscape geometry I'd noticed before London's Earthstars began to occupy my attention. However, while I was puzzling over London's sacred geometry, at least two other people seem to have been finding similar and somewhat simpler, patterns elsewhere in the world.

THE CIRCLE OF PERPETUAL CHOIRS

In late 1983, Earthquest News magazine carried an article entitled The Circle of Perpetual Choirs. A group of researchers, including John Merron and Alan Cleaver, had been tracing an ancient pattern on the landscape originally identified by John Michell from references in the Welsh Bardic Triads. It was a circle of ten sites where ancient choirs were said to have maintained a ceaseless chant, 24 hours a day, every day; hence the name.

John Michell's references to The Circle of Perpetual Choirs can be found in his book, CITY OF REVELATION. They show an arc of the circle between four of the known choirs; Goring on Thames, Stonehenge, Glastonbury and Llantwit Major on the Welsh Coast. John Merron had progressed the geometry to complete the circle and to investigate whether anything of significance was to be found at the places subsequently identified. As a landscape pattern, the circle of ten sites bore a remarkable similarity to the initial ten point star I had found on the London landscape, although on a much larger scale. John Michell's calculations put the distance between points of the decagon on the Perpetual Choirs' circle at 39.6 miles or 316.8 furlongs.

Apart from the obvious similarity in appearance, there is a startling correspondence between John's decagon of choirs and the Earthstars decagon. The length of the Earthstars' decagon sides are exactly five miles each. In inches that is 316,800. The Perpetual Choir's decagon sides are 316.8 furlongs, a remarkable repetition of the figure which surely cannot be a coincidence. Even more remarkable is the factor which distinguishes the difference between these two figures and between furlongs and inches. The number of inches in a furlong is 7,920, precisely the same number of miles in the Earth's mean diameter.

The perimeters of the decagons would therefore be 31,680 furlongs (perpetual choirs) and 3,168,000 inches (Earthstars) respectively.

The two designs are clearly constructed on the same synergistic principles. Given these proportions, it is easy to calculate the difference in size between the two decagons.

Significantly, the Circle of Perpetual Choirs is exactly 7.92 times larger than the corresponding Earthstars circle. At the risk of becoming repetitive, I have to point out that the number 792 relates to the mean diameter of the Earth as 7,920 miles. This cannot be a coincidence.

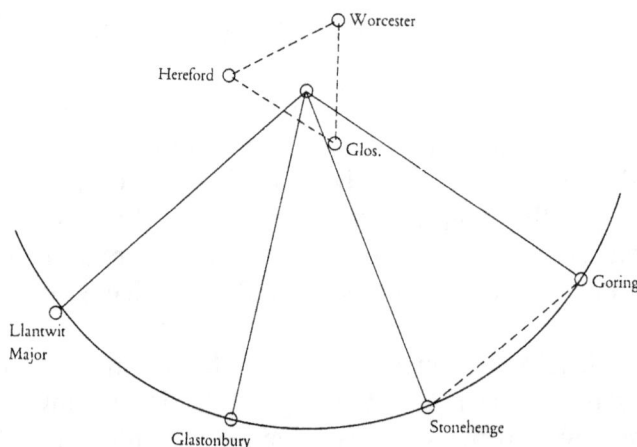

Illustration 142: The arc of the Circle of Perpetual Choirs first illustrated in John Michell's book CITY OF REVELATION (reproduced by permission of the author).

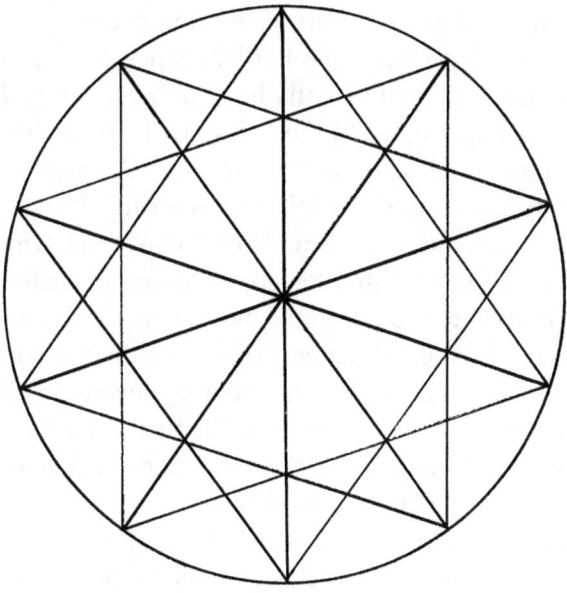

Illustration 143: The Circle of Perpetual Choirs illustrated in Earthquest News was a circle of ten sites similar to one of the Earthstars' patterns.

The parallels between their dimensions may result from their common use of a system of proportional harmony with the planet. Then again, the existence of the Circle of Perpetual Choirs does suggest it was designed and marked with sacred sites in a distant era when there was a greater awareness of phenomena such as the Earthstars landscape geometry. Certainly, it seems to indicate that the Circle of Perpetual Choirs and the Earthstars patterns may be related phenomena.

What is more certain is that the standard British units of inches, feet, yards, furlongs and miles were clearly conceived in as part of the same system of proportional harmony resonant with planetary dimensions.

With this in mind, I think it is relevant to point out that other British systems of measurement also reflect the Earthstars numbers.

Our old UK monetary system, for instance, had units of twenty made up of four fives like the Earthstars circle. There were four crowns in a pound, each crown being five shillings making twenty in all. Incidentally, the druids are known to have calculated in fives and twenties too, as did the Mayans. Another monetary number was 12, the number of pennies in a shilling and months in a year.

Our time scale also has parallels with the Earthstars patterns.

Double the 30-point star to sixty and it becomes the basis of our clock face, which is also divided into sections of five and twelve.

In both Stonehenge and the Earthstars design, the 30-point star circle exhibits clues that it should actually be doubled to sixty. The Earthstars 30-point is only directly aligned to the N-S axis. To be aligned to the E-W axis as well, it would need to be a 60-point star. At Stonehenge there would have been sixty stones in the original sarsen ring, 30 upright and 30 lintels on top of them. Do these patterns bear comparison with our divisions of time, sixty seconds in each minute, sixty minutes in each hour ?

Time, of course, is determined, not by the movement of hands around a clock face, but by the movements of the Earth as it spins around the sun and rotates on its own axis. We know that the geometry of Stonehenge was dictated by the movements of the Sun, Moon and Earth, so once again, this may be another indication that the Earthstars design is a structure built upon hidden cycles of planetary energies.

Like the Circle of Perpetual Choirs material, the great majority of new information on this kind of subject finds its way into print via small circulation magazines like Earthquest News. Usually, they're produced by groups of enthusiastic and dedicated researchers who are delving into their local mysteries. I subscribed to several of them as they contained an enormous amount of useful background information on the subject of leys, ancient sites, folk-lore and Earth mysteries in general.

I was particularly interested in Earthquest News because Andrew Collins, who produced it, had a unique approach to these matters for which he had coined the term Psychic Questing.

Basically, it involved intuitive interaction with ancient sites, their spirit guardians or spirit of place and often, surprisingly, resulted in verifiable facts, useful information or in many instances, tangible artefacts being unearthed.

Obviously, this is an area of research which attracts enormous scepticism, but it fascinated me because it entailed the kind of psychic experiences which led me to the Earthstars discovery. These were just the sort of things that were reported regularly in the pages of Earthquest News by Andy, members of his Psychic Questing Group and contributors to his magazine.

NB. Andrew Collins no longer produces Earthquest News as a magazine, having moved on to writing weightier volumes such as THE SEVENTH SWORD, THE BLACK ALCHEMIST, FROM THE ASHES OF ANGELS, THE SECOND COMING and THE GODS OF EDEN.

GENISIS

A second example of star-based landscape geometry came into the public arena with the launch of David Wood's astonishing book GENISIS in 1984. This featured a set of intriguing geometric patterns around the Rennes Le Chateau area of South East France.

The geometry was less regular than London's Earthstars and on a smaller scale, but it was landscape temple geometry nonetheless and like my Earthstars, included a design which merged pentagonal and hexagonal symmetry. In this case, through the curious figures of a distorted pentagram and hexagram interlocked on the Languedoc landscape and defined by a variety of ancient churches, castles, natural peaks, plus the odd chateau.

Unlike London's Earthstars, this discovery was not the result of any subjective psychic escapades. David Wood is qualified in the fields of trigonometric and topographical surveying, as well as cartographical reproduction. These are the skills which led to his revelations and his ability to verify the accuracy of his landscape geometry far exceeds mine.

Nevertheless, there was an intriguing parallel with my psychic experiences. As the title GENISIS suggests, David Wood firmly linked these mysteries with the cult of Isis, a deity who I had already associated with the White Goddess apparitions that had initiated the Earthstars discovery. His conclusion was that a highly complex geometric plan governed the placement of various important structures in the Rennes le Chateau landscape and that the ingenuity of the work implied an intelligence beyond the known abilities of our ancestors, possibly of extra-terrestrial origin.

Prior to the publication of my first book, Earthstars, in 1990, these three examples were all I had for comparison. Since then, examples of the enigma have escalated and a remarkable number of other patterns have come to light, including Peter Knight's thoroughly researched Wessex Astrum in 2009.

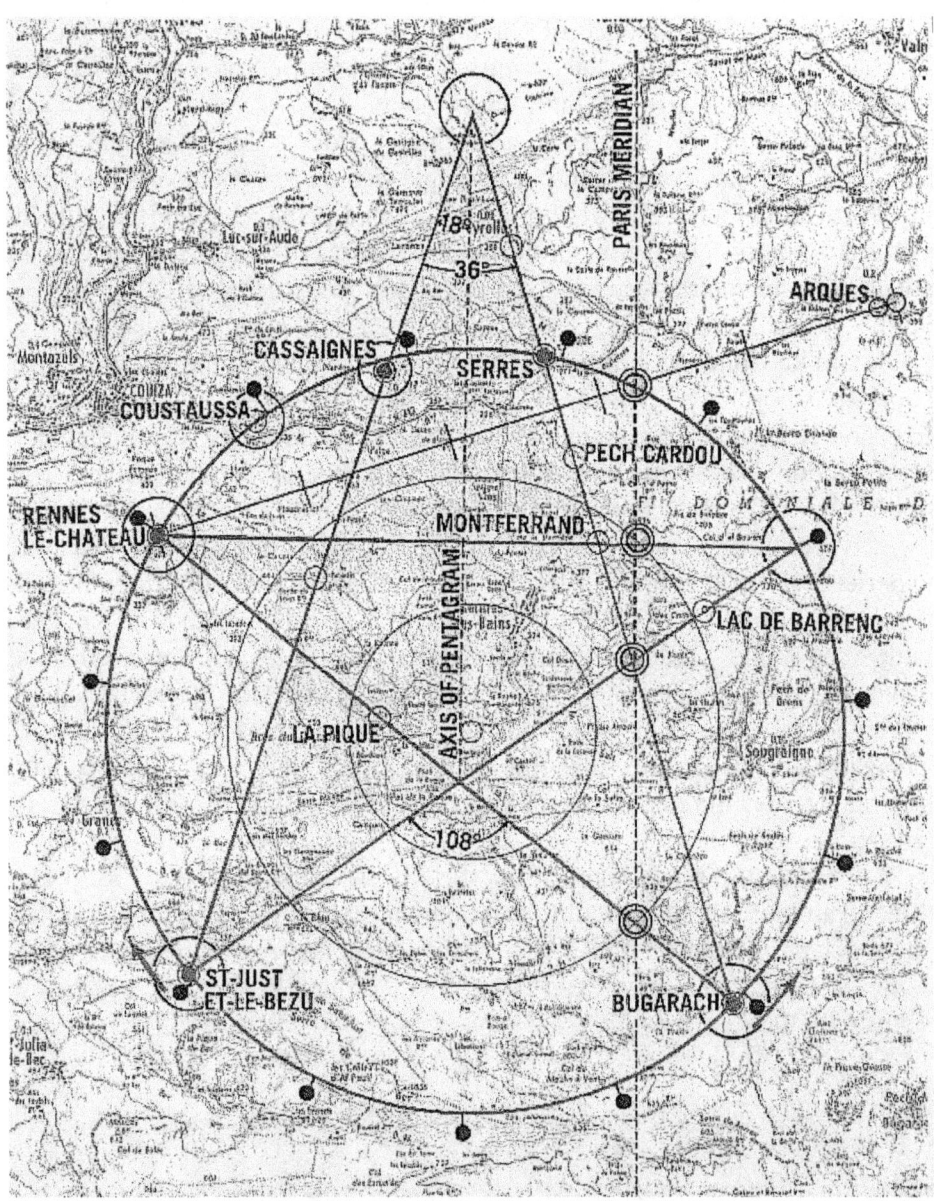

Illustration 144: One of the star patterns on the landscape around Rennes Le Chateau, which form the basis of David Wood's books GENISIS and GENESET.

THE HOLY PLACE

In 1991, Henry Lincoln's book, THE HOLY PLACE, expanded on the geometry of the Rennes Le Chateau area. From this I learned that David Wood's starting point had been a perfect pentagram of natural peaks initially identified by Henry Lincoln whilst investigating the mysteries of Rennes Le Chateau. (These are far too complex to adequately summarise here and form the basis of such books as THE HOLY BLOOD AND THE HOLY GRAIL and THE MESSIANIC LEGACY)

Obviously a pentagram of natural peaks doesn't need to have been consciously laid out by anyone other than nature, however, Henry Lincoln's evidence does suggest that an awareness of the landscape pattern was an important consideration in the mysteries surrounding Rennes le Chateau. Such a pattern would have marked this area as a place of immense natural sanctity but does raise the question of how it came to be noticed in an era when there were no large scale maps, aerial photography or air travel.

The pentagram must have been noticed and its significance understood because, according to Henry Lincoln, in consequence, it has been incorporated into a vast geometric temple ground plan covering an extensive area. Many of the most important structures of the region had been carefully located on key positions within it. This grand design incorporated some of David Wood's geometry as well as that discovered by Henry Lincoln himself.

Points of comparison with London's Earthstars include not just a perfect pentagram, but also a perfect ten-point star, albeit at a smaller scale, a diameter of just under six miles as opposed to London's 16.2. To underline the possible connection with the Earthstars decagon and John Michell's Circle of Perpetual Choirs decagon, the number 3,168 turns up again. Six miles is precisely 31,680 feet.

Despite the fact that he has a strong preference for hard, objective evidence based on fact, some of Lincoln's conclusions bear comparison to my own intuitive impressions (and the reasoned conclusions of John Michell and David Wood) that these phenomena are fundamentally associated with a goddess-focused religion. At the end of Chapter Seven, he states categorically: " **For an earlier culture, Rennes le Chateau was a gigantic, god-given temple to the Mother Goddess.** "

Rennes le Chateau does, after all, translate as "the house of the Queen."

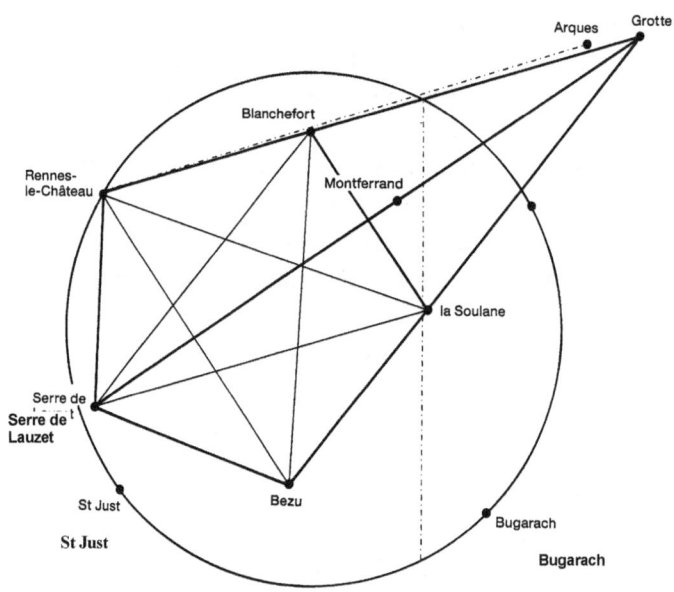

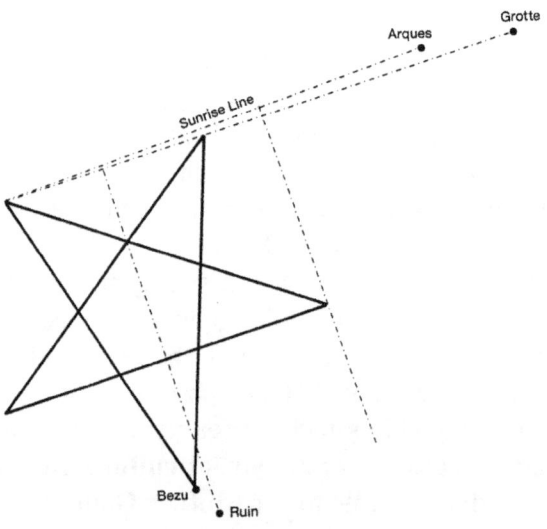

Illustration 145 and 146: Diagrams from Henry Lincoln's fascinating book THE HOLY PLACE, showing the pentagonal arrangement of sites in SW France which has an identical sunrise alignment to the London Earthstars' pentagram.

PI IN THE SKY

In 1992 a fourth example of star-based landscape geometry emerged. Rider published a book entitled PI IN THE SKY, the result of Michael Poynder's life-long study of the ancient wisdom tradition, encoded by Stone Age Man in monuments, rock carvings and the design of jewelry artifacts.

Their geometry reveals a surprising awareness of archetypal structure and its associations with the life force which pervades the universe. The application of such knowledge would have been designed to contain, manipulate and direct that force. As it says in Michael Poynder's cover notes:

"Their sacred sites modeled and put to work the geometry of the solar system, creating extraordinarily powerful concentrations of the vital force."

In the book, the late Clive Beadon, a prominent member of the British Society of Dowsers, is quoted as saying:

"The surface of the Earth is covered by an interlocking pattern of energy lines, the nature of which is not fully understood."

Amongst the patterns to which he refers are some he describes as Earth Stars, discovered by himself and by Michael Poynder through the art of dowsing. They share some similarities to the Earthstars super-imposed upon the London landscape, but are principally six-point stars rather than the predominantly pentagonal blend of complex symmetries I had found.

However, Michael Poynder believes, as I do, that these geometric patterns are carriers of a vital-life force, although he would add the proviso that lines on a map, in themselves, would not be indicative of energy flows in the landscape, unless confirmed as such by dowsing.

I would agree with him, although, having felt the energy myself at many of the places associated with the Earthstars patterns, I suspect dowsing could turn up a yet another level of energy patterns in the Earthstars sites.

In recent years, some well-known authorities on Earth Mysteries have exhibited a tendency to disregard the findings of dowsers as too subjective to be admissible. Their current stance on the matter seems to be that lines of energy across the land do not exist; That any linear alignments are actually

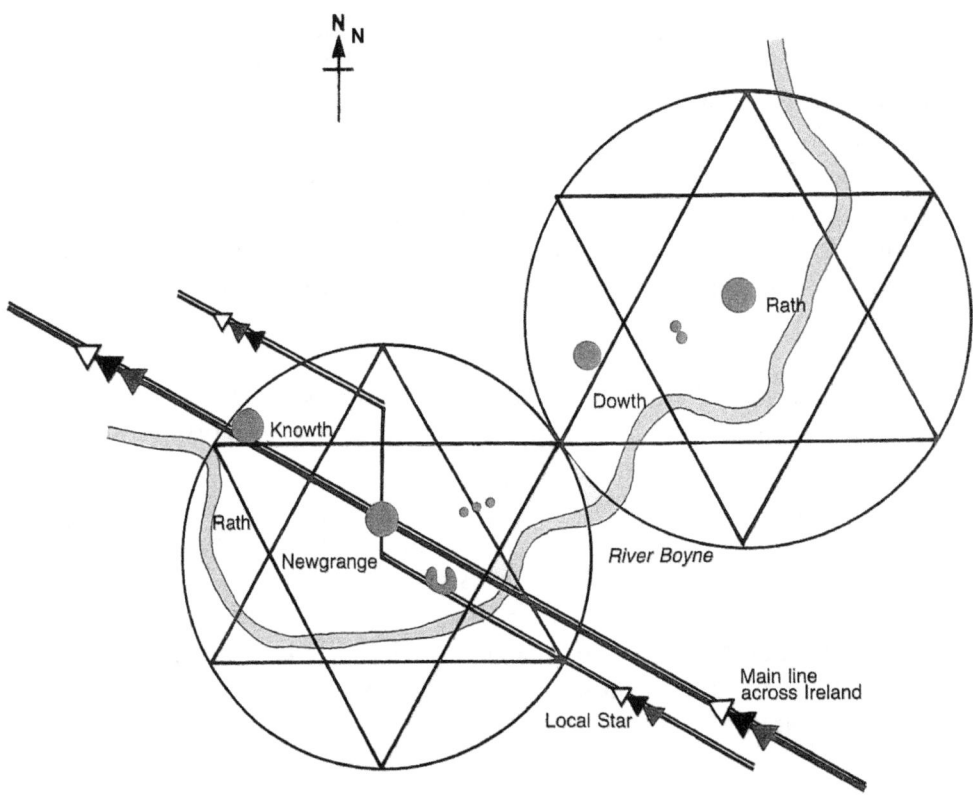

Illustration 147: Examples of Michael Poynder's dowsed Earth Stars. These energy patterns were dowsed in the Newgrange area of Ireland, (reproduced from PI IN THE SKY with permission of the author).

something else. Shamanic flight paths or spirit lines are two suggestions presently being promoted by the anti-dowsing lobby. Their inflexible stance ignores the basic fact that dowsers do actually detect something and that most of them regard energy and spirit as two sides of the same coin. Almost without exception dowsers will tell you that the energy they are detecting has a spiritual basis.

Michael Poynder, in this respect, is no exception.

He demonstrates a clear awareness that this phenomena is linked to our spiritual dimensions and to the life-force which continues to work through our sacred sites, particularly those built upon the ceremonial places of our ancestors who, it would appear, understood these matters far better than modern man.

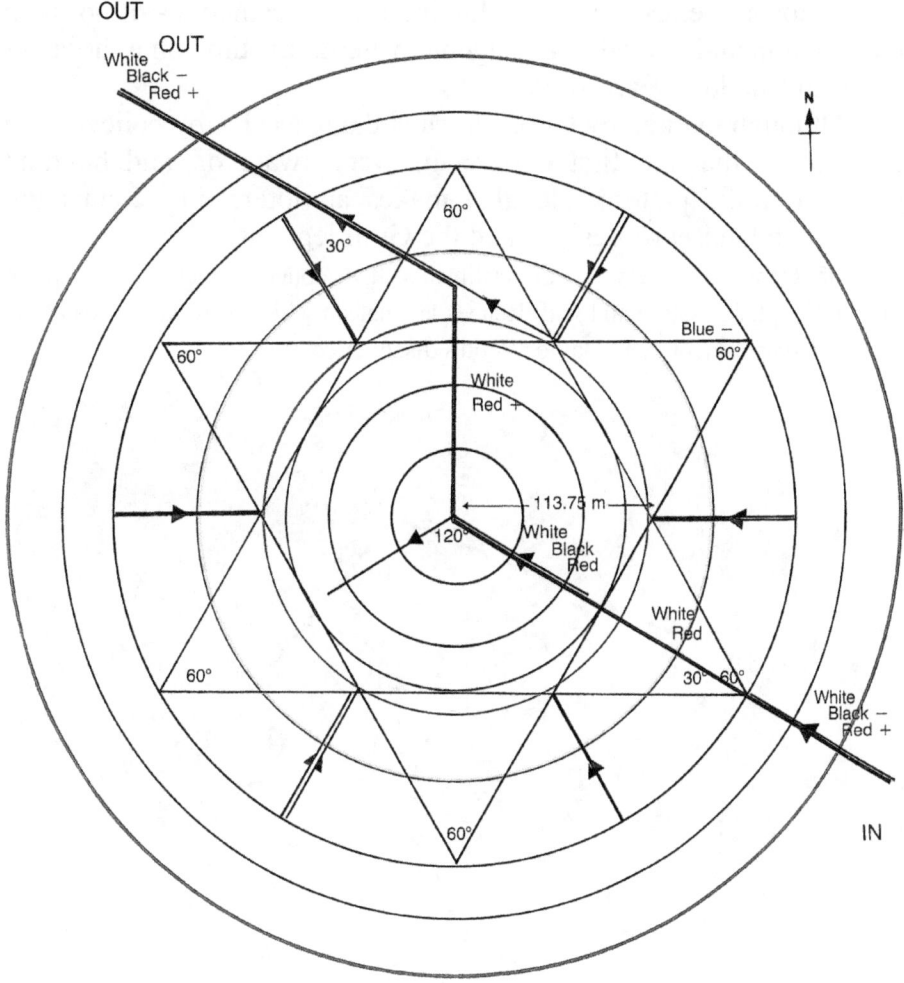

Illustration 148: The basic diviners' Earth Star pattern which Clive Beadon and Michael Poynder detected in many locations (reproduced from PI IN THE SKY by kind permission of the author).

ON EARTH AS IT IS IN HEAVEN

In 1996, ON EARTH AS IT IS IN HEAVEN, appeared from Rhaedus Publications, written by Greg Rigby and sub-titled REVELATIONS OF FRENCH CATHEDRAL LOCATIONS.

It shows a series of patterns linking the key cathedrals of northern France and includes another pentagonal pattern, this time on a far larger scale than London's Earthstars.

Although marked by Christian cathedrals, Mr Rigby believes and demonstrates that the first Cistercians were aware of, and building upon, a Druidic pattern. He also makes an interesting connection between the landscape designs and the Grail legends.

Mr Rigby's discovery has direct links with Avebury, Stonehenge, Philae in southern Egypt, Jerusalem and with Rennes le Chateau which, of course, lies at the centre of Henry Lincoln and David Wood's discoveries.

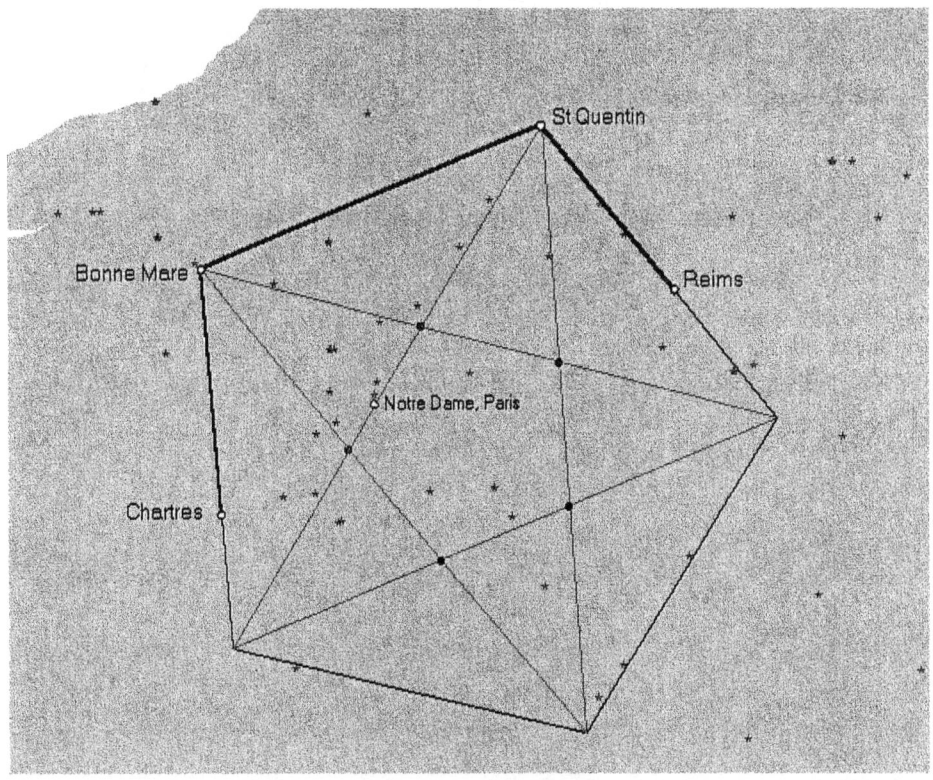

Illustration 149: The large scale pentagram of sites featured in Greg Rigby's book.

THE KEYS TO THE TEMPLE

Another 'must-read' book on this subject is THE KEYS TO TEMPLE (1997) by Henry Lincoln.

It's an entertaining update of his Rennes Le Chateau discoveries which includes some details of similar star patterns found in Brittany, in Norway and at Bornholm in Denmark, either by people who had been inspired by Henry's original book, or who had stumbled upon the discoveries independently.

OTHER EARTHSTARS PATTERNS

Since the publication of my first "Earthstars" book in 1990, I have been in touch with people who have discovered star patterns overlaid on a variety of places, including Majorca, Tunbridge Wells, Stonehenge, Avebury, Birmingham, Oxford, Manchester and parts of Oxfordshire and Derbyshire. Needless to say, I have extended the Earthstars alignments from London to the rest of the country and have also found more of these patterns myself.

Harrow on the Hill, for instance, is the centre of a very interesting pentagram, defined by its surrounding hills, including Horsenden Hill in Greenford, Barn Hill at Wembley and Belmont in Stanmore. The fifth point is less obvious, but may either be the summit of Dabbs Hill in South Ruislip or a high point nearby in Eastcote Lane.

This pentagram is by no means perfect. The 72 degree angles from St. Mary's Church at Harrow-on-the-Hill are accurate, but its five points are different distances from the centre, giving rise to an irregular five-point-star. Nevertheless, it is linked into the larger Earthstars geometry. It shares its Horsenden Hill point with the Earthstars' pentagram and the Horsenden to Barn Hill side of the figure is close to the midsummer sunrise alignment from Horsenden Hill to St. Mary's in East Barnet which, as you may recall, passes over Barn Hill.

There are also several fascinating correspondences between Harrow-on-the-Hill and Rennes le Chateau. Both are hilltop locations with a notable St. Mary's dedication to their local church. They are both of sufficient antiquity to have been places dedicated to a divine mother prior to the advent of Christianity; and are both surrounded by hills which form a natural pentagram.

I wouldn't be at all surprised to find that Harrow-on-the-Hill's peculiar landscape geometry conceals almost as many mysteries as Rennes'.

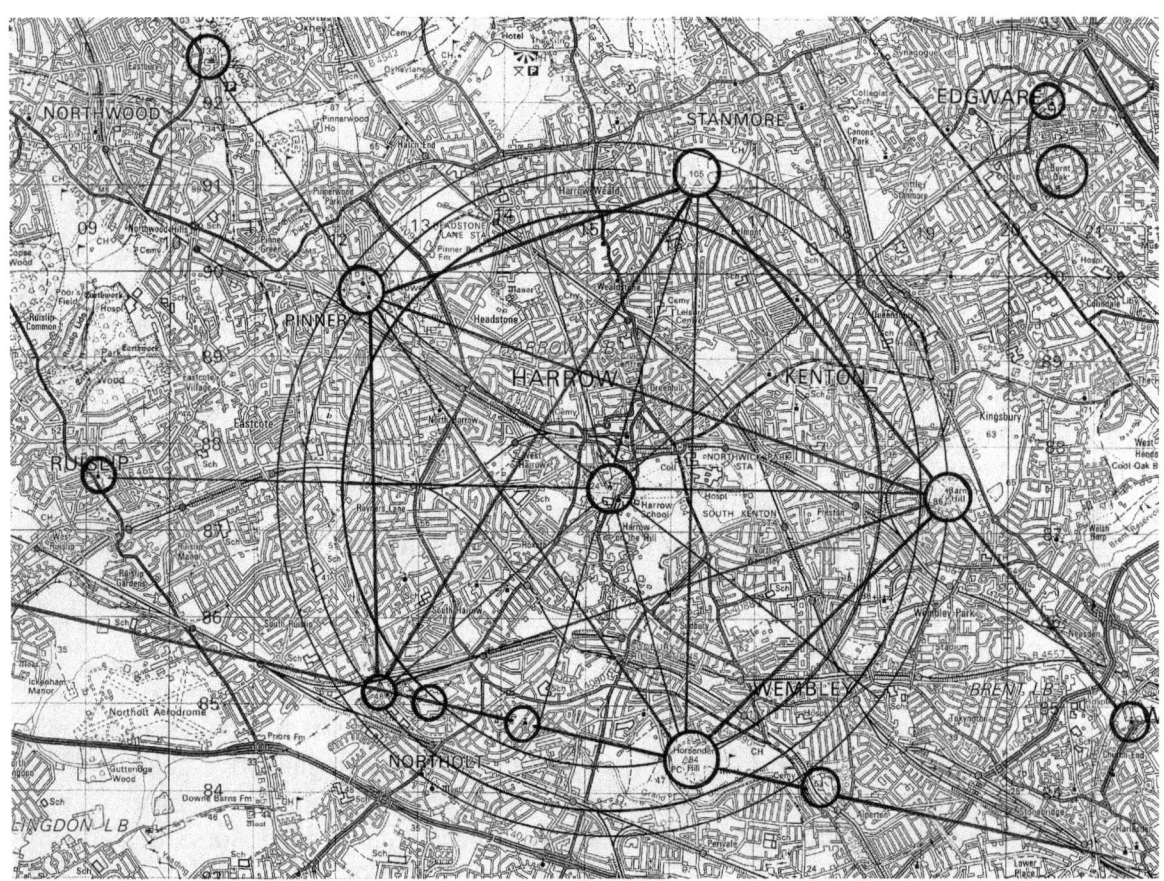

Illustration 150: Map of a slightly irregular pentagram centred on Harrow-on-the-Hill, marked by prominent hilltop sites; Horsenden Hill at Greenford, Barn Hill at Wembley, Belmont at Stanmore and Dabbs Hill in South Ruislip. O.S. map © Crown Copyright. All rights reserved. License No. 100029558.

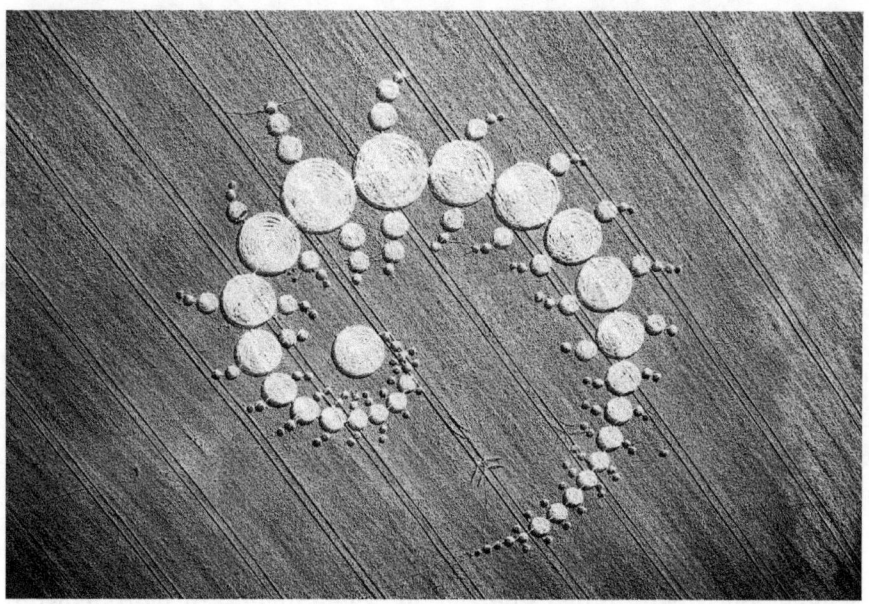

Illustration 151: The complex crop pattern which appeared near Stonehenge in 1996. It contains 151 circles and formed within the space of 15 minutes.

Illustrations 152 and 153: Many crop circle formations have an obvious geometric basis. These two appeared in 1998, bearing a similarity to the 10-point star of the Earthstars' design.

THE SACRED GEOMETRY OF CROP CIRCLES

In the context of Earthstar patterns, there is, of course, the phenomenon of crop circles to be considered.

Despite the suspicious attempt some years ago to de-bunk the subject and attribute the entire enigma to two hoaxers with nothing better to do after the pubs had shut, the circles continue to appear annually and defy explanation.

The complex formation which sprang up 300 yards from Stonehenge in July 1996 is an excellent example. It covered an area 500ft by 900ft, contained 151 interlaced circles of various sizes and, according to reliable witnesses, formed in the space of fifteen minutes in broad daylight. Most people couldn't have drawn it accurately on a piece of paper during that time, let alone marked it out precisely at this scale in a field.

By contrast, the hoaxers who claim to be responsible for the crop of circles have consistently failed to create the same quality of workmanship in front of observers, even when allowed several hours to complete their labours.

Most genuine crop circles have been clearly based on precise geometric principles; Star-shaped formations are common, with hexagonal patterns appearing regularly. Since 1998/99, several five and ten-point stars have cropped up.

Dowsers investigating this phenomenon maintain that genuine crop circles (as opposed to the hoaxers' clumsy imitations) invariably form on dowsable leys, hinting that their mysterious energies may somehow be employed in the circles' formation.

My own suggestion is that what we may be witnessing here is a vertical component of the Earth's energy lines; one which connects our planet's intrinsic energies to the rest of the cosmos and which may be a natural effect of the Earth's journey through the universe.

Since the energy of the Universal Life Force seems to work through these geometric patterns, it should not be at all surprising that a sudden powerful pulse may be sufficient to imprint them on a field of crops.

The fact that the crop circles have increased in number and in geometric complexity over the years seems to indicate that the Earth is entering a phase where it is subject to increasingly powerful impulses from elsewhere in the universe.

JOHN MICHELL'S SACRED GEOMETRY OF JERUSALEM

Several more examples of sacred geometry deliberately employed in the landscape appear in the works of John Michell. This is perhaps predictable since he is the world's leading authority on such matters. His various books demonstrate an awesome knowledge of esoteric mathematics, numerology, metrology, sacred geometry, ancient cosmologies and geosophy

AT THE CENTRE OF THE WORLD, published by Thames and Hudson in 1994, includes a series of maps originally included in TWELVE TRIBE NATIONS, illustrating how the original provinces of Ireland were based on a pentagonal division of the island.

A far more impressive example features in the later chapters of TWELVE TRIBE NATIONS AND THE SCIENCE OF ENCHANTING THE LANDSCAPE co-written with Christine Rhone, (Thames and Hudson 1991). Hidden in the foundational plan of Jerusalem, John Michell unearthed a pattern of interlocking pentagrams, forming the basis of temple ground plan and with some of the most important holy places in the world defining its key locations.

The most obvious figure in Jerusalem's sacred geometry is a pentagon centred on an alignment John Michell refers to as the Messianic Ley, which he states was first identified by a Dr. Asher Kaufman. This forms the main east-west axis of the city and passes from the summit of the Mount of Olives, through the Golden Gate, over the rock of the former temple (marked by the Dome of the Spirits) and on to the hill of Cavalry or Golgotha beneath the church of the Holy Sepulchre. It then continues to the western tip of the old city walls.

This pentagon would seem to have been deliberately planned and laid out. Its central axis is the Messianic Ley alignment. Three of its lines are marked by streets dating back to the Roman period or possibly earlier, whilst its eastern side defines the west end of the former temple. The scale of this impressive figure is much smaller than London's five-pointed star. The pentagon's sides are approximately one quarter of a mile each, making the total perimeter only a mile and a quarter.

Not surprisingly, these proportions reveal some of the usual numbers. A mile and a quarter is 79,200 inches.

This pentagram is linked to second, reciprocal pentagram which overlaps the first. Both have their shared arm defined by the Cardo Maximus, the main street of the city, established by the Roman augers as Jerusalem's main north-south axis. A series of reducing pentagrams

London   City of Revelation

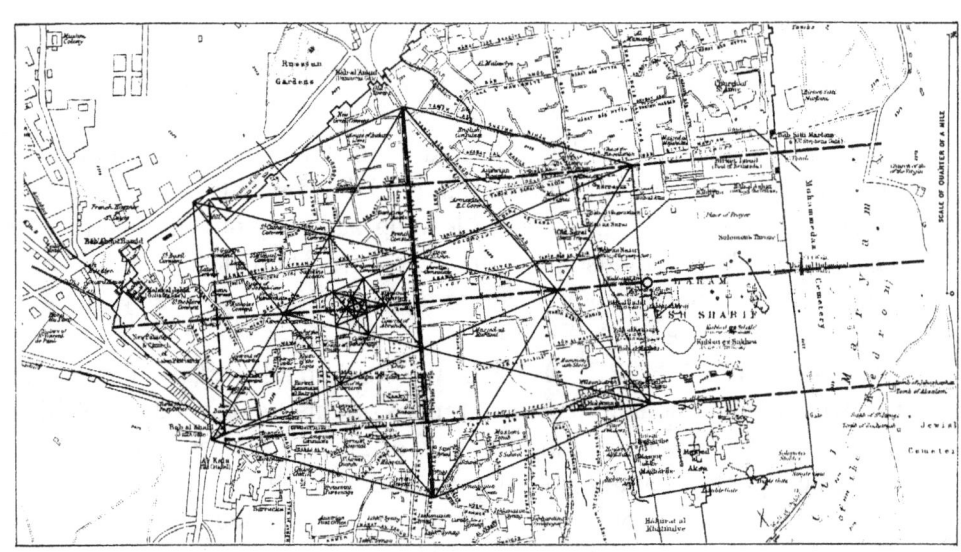

Illustration 154: One of John Michell's maps showing the sacred geometry underlying Jerusalem, reproduced with the author's permission from TWELVE TRIBE NATIONS AND THE SCIENCE OF ENCHANTING THE LANDSCAPE.

within this second pentagram proves that its exact centre lies upon the rock of Golgotha. This, apparently, is not the full extent of the grand design, for as John Michell explains, there is far more:

"This lovely pentagonal scheme, forming a six-sided, equilateral figure, provides the framework of the symbolic design which Hadrian's Roman augers planned, or more probably, renewed over Jerusalem. There is evidence of other geometrical patterns linked to it and to the walls and streets of the old city. Possibly the full scheme was a synthesis of all geometric types, hexagons, heptagons, octagons and so on to create a geometer's image of the universe."

A common theme running through much of John Michell's work is that an ancient code of knowledge was once employed to produce harmony between humanity and nature and that it is based on a numerical cosmology or canon of proportion which was at the root of all ancient cultures.

" In the known civilisations of antiquity, China, Babylon, and Egypt, the canon of number was venerated as the source of all knowledge and a guide to rightful conduct. "

"One numerical code has fashioned the whole of ancient mathematics, music, astronomy, chronology, metrology and every variety of craft. It has left its mark on every relic and tradition of ancient cultures. There is nothing artificial about it, for the conclusion of these researches is that the various orders of natural phenomena do indeed conform to certain similar patterns of number."

"The ancient philosophers understood the structure of number to be analogous with the structure of creation."

These quotations are all taken from John Michell's book THE DIMENSIONS OF PARADISE published by Thames and Hudson in 1988. It presents a vast amount of evidence in support of the viewpoint that harmonious numerical standards, which Michell calls The Dimensions of Paradise, underlie the created world and that this numerical canon was employed, not only in the proportions of sacred structures, but in the establishment of ritualised, sacred landscapes.

Even more material on these themes appeared in John Michell's AT THE CENTRE OF THE WORLD in 1994. The entire book expounds the geomantic practice of establishing a centre of power or "fixing the dragon" which was explained briefly in the last chapter. It demonstrates that this concept of a central cosmic axis of power was the basis of geomantic systems throughout Europe in the distant past and, as well as being employed in the establishment of sacred sites, was also used on a wider scale to create ritual land divisions.

THERE'S MORE THAN ONE PENTAGON IN WASHINGTON

It will come as a surprise to most people to find that a relatively modern city has been laid out incorporating many of these ancient principles, but a street plan of the U.S. capital provides instant evidence that it has. A pentagonal design is clearly delineated by the city's streets. Its focal point is the White House and its central north-south axis is marked by 16th Street, the path through La Fayette Park, the White House, the Zero Milestone, Haupt Fountains and the Jefferson Monument. This is not a regular pentagram and forms part of a much larger and more complex plan, a pattern of power on the landscape reflecting the power and influence of the U.S. government. The angles involved suggest an ad quadratum basis initially, although though the original plan of the city is based on a huge pentagram.

Having said that, one or two things are pretty obvious. An east-west alignment from the Lincoln Memorial to the U.S. capital building seems an important foundation of the plan since the Washington Monument has been located on this instead of the N-S White House alignment. Also, one axis of the Pentagon building is aligned directly towards the White House.

What sets these patterns and alignments apart from more ancient examples like London's Earthstars is that we know who was responsible for them. The layout of Washington was based on plans prepared by Major Pierre Charles L'Enfant, a French Freemason, under the direction of George Washington, Thomas Jefferson and other eminent U.S. statesmen. Clearly the Masonic traditions of this era still embodied some ancient wisdom and understanding of this kind of landscape engineering.

The Sacred Geometry of Washington DC is revealed in considerable detail by Nicholas Mann in his book of that title.

London   City of Revelation

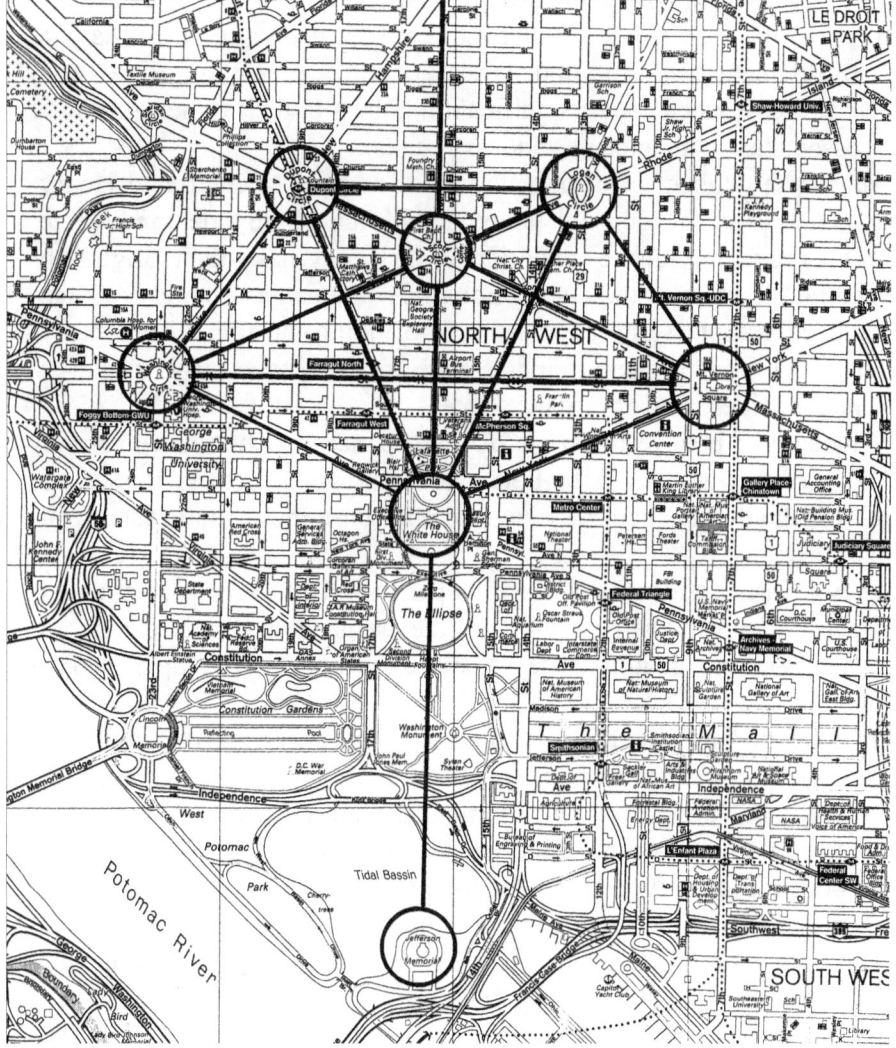

Illustration 155:
A pentagram in Washington's street plan points directly to The White House.

I would hesitate to claim all these examples have the same basis, since a far more detailed study of all of them would be necessary before forming any definite conclusions. At the very least, they prove that the Earthstar patterns I found over London are not an isolated phenomenon and that there is supportive evidence for them elsewhere.

As a general theme, geometric star patterns found in the landscape seem to be on the increase, and several of them, it has to be noted, have a variety of elements in common with London's Earthstars discovery; particularly, the repeated use of significant numbers in their dimensions and a strong association with ancient goddess figures.

Out of the published material, the work of John Michell, and Henry Lincoln, as well as the dowsed patterns of Michael Poynder and Clive Beadon, all provide evidence of designs which may have their origins in nature and the structure of the planet. Equally, some patterns could not have been achieved without some degree of pre-planning, or embellishment by human hand. This would seem to indicate we are re-discovering evidence of an ancient and widespread spiritual science.

Personally, I believe the basis of these patterns to be totally natural expressions of a Universal Life Force and that our ancestors possessed faculties which allowed them to recognise and sense these subtle forces, in a number of ways.

They invariably identify a place as particularly sacred. Any obvious pattern detected could then be used as the basis of a grand scheme to sanctify a larger area, creating a sacred landscape based on a temple ground plan.

In London, I believe the vast scale of such an endeavour makes it likely that these particular patterns are a completely natural phenomenon, merely marked in places by human additions to the landscape. This is not to say that the patterns found by John Michell in Jerusalem, for instance, or David Wood in France were not deliberately designed. Their scale makes for a far more manageable project than London's huge Earthstars mandala.

The assumption that London's Earthstar patterns are natural means that they represent a complex circuit diagram in the Earth's life force. They are a temple design proven to have been used in the construction of the Great Pyramid and Stonehenge and one whose proportions are resonant with both the structure of the planet and cosmic harmonies.

It adds up to an extremely significant place.

Possibly a holy city.

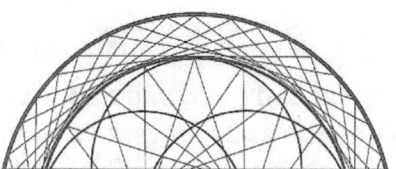

Chapter Thirteen
The City of Revelation

Anyone who is at all familiar with the many facets of dear old London will almost certainly find it difficult to think of the place as a Holy City. After all, it does have more than its fair share of areas where the impression that you are standing on hallowed ground is likely to be the last thing that hits you. Nevertheless, it has the sacred geometry of a holy city, discernible in the relative locations of some of its most ancient sites.

On its own that would be astonishing enough, but there is more. London's sacred geometry has some very curious connections to another Holy City; the New Jerusalem; the celestial city seen in a vision by St. John The Divine and recorded in The Book of Revelation.

> "And I saw a new heaven and a new earth; for the first heaven and the first earth were passed away and there was no more sea. And I John saw the Holy City, New Jerusalem, coming down from God out of Heaven, prepared as a bride adorned for her husband."
>
> Revelation 21; vs 1-2

A few verses later, he begins to describe the city.

> "And he (the angel) carried me away in the spirit to a great and high mountain, and shewed me that great city, the holy Jerusalem, descending out of heaven from God
>
> Having the glory of God; and her light was like unto a stone most precious, even like a jasper stone, as clear as crystal."

> **And had a wall great and high, and had twelve gates and at the gates, twelve angels and names written thereon, which are the names of the twelve tribes of the children of Israel;**
>
> **On the East three gates; on the north three gates; on the south three gates; and on the west three gates, And the wall of the city had twelve foundations, and in them the names of the twelve apostles of the lamb."**

Revelations 21; vs 10-14.

It doesn't sound like London, does it? No mention of Big Ben, Tower Bridge, the Statue of Eros at Piccadilly Circus, Nelson's Column, no monstrous gherkin or giant Canary Wharf Obelisk.

The City of Revelation is no ordinary city. If it exists at all, it is understood as a vision and an allegoric description of the perfect state or the Kingdom of Heaven. It is therefore a reference to our spiritual dimensions rather than the physical plane and like the spiritual dimensions of our temples and cathedrals, its construction may be represented by a plan of sacred geometry.

To alert us to this fact, the various descriptions of the city given in The Book of Revelations, draw particular attention to its proportions and from these, the sacred geometry may be deduced.

> **" And he (an angel) that talked with me had a golden reed to measure the city thereof, and the gates thereof, and the wall thereof.**
>
> **And the city lieth foursquare, and the length is as large as the breadth; and he measured the city with the reed, twelve thousand furlongs. The length and breadth were equal.**
>
> **And he measured the wall thereof, an hundred and forty and four cubits, according to the measure of a man, that is, of the angel."**

Revelations 21; vs 15-17.

John Michell is probably the world's leading scholars of ancient metrology, sacred geometry and related subjects. In his books, CITY OF

REVELATION and THE DIMENSIONS OF PARADISE, he examines, analyses and explains the significance of the New Jerusalem in great depth. Amongst other things, he reveals that the proportions and measures of the New Jerusalem have hidden meaning for those well versed in the arcane knowledge of sacred geometry.

Like most things esoteric, the dimensions given make no sense to the uninitiated. In terms of scale, 12,000 furlongs and 144 cubits could not possible relate to the same physical wall. A furlong is 660 ft, whilst a cubit of any sort, Egyptian, Sumerian, Roman or whatever, is less than 2 feet. 12,000 furlongs would make the city walls a massive 1500 miles in length, breadth and height. On the other hand, calculations based on an Egyptian cubit of 1.728 feet would give a wall of just 248.832 feet, which is about 82 yards and less than half a furlong.

The explanation for this major discrepancy, according to John Michell, is that the measures so obviously reflect opposite ends of a scale that their incompatibility itself is significant. It emphasises, he suggests, that the New Jerusalem;

"represents both the macrocosm and microcosm, the order of the heavens and the constitution of human nature."

This may possibly be hinted at in the curious phrase;

"According to the measure of a man, that is, of the angel."

Where the measures of angels and men share common ground is in the divine proportions of sacred geometry and their underlying mathematical principles. If we ignore the relative scales of furlongs and cubits for a moment and, reading between the lines, that is exactly what we are being urged to do, we find that these dimensions have immense significance purely in their numerical relationship.

Take 12,000 furlongs, for instance. In feet, that would be 7,920,000. 7,920 is a number you might recognise from a previous chapter. Certainly those aware of its significance would have recognised it immediately. It is the mean diameter of the Earth in miles.

Equally, 248.832 feet is not simply 144 Egyptian cubits. Move the decimal point two places to the right and it becomes another measure of planetary significance. 24,883.2 in miles is the Earth's mean circumference. So here we have two indirect, yet unmistakable references to planetary proportions, in a document supposedly written almost two thousand years ago.

That in itself raises a number of important questions, but constraints of time and space mean they will have to remain unanswered for a while longer.

Whatever its scale, the sacred geometry of the New Jerusalem is clearly based upon the proportions of the planet as well as those of an ideal celestial city. Its foundation therefore is related directly to the sphere of the Earth and this is clearly demonstrated in the proportions of the two figures.

24,883.2 divided by 7,920 produces a revealing 3.1418. This is remarkably close to the numerical statement of Pi (3.14159) vital to the formulas for calculating the area or circumference of a circle from a given radius. Pi is generally taken to three decimal points and used as 3.142. Naturally, 3.1418 can be reduced to three decimal places in the same way to become an identical 3.142.

From these numbers it would have been obvious that reference was being made to the relative proportions of a circle or sphere, as well as the dimensions of the Earth. Yet the New Jerusalem is also referred to in no uncertain terms as foursquare, hinting that this may be understood as a clue that we are dealing with three dimensional geometry and the symbolic concept of the cube or square of the Earth.

What we are being gently steered towards here is a mystical union of square and circle, cube and sphere, which also incorporates the proportions of the planet. Only one construction in sacred geometry fits this description. As John Michell rightly points out in CITY OF REVELATION, it can only be a reference to the squared circle, or its close relative, the cubed sphere.

The biblical measures, however, also contain a further clue which indicates this puzzle has its basis in sacred geometry; 144 cubits is 12 x 12. Its equivalent in inches, 248,832, is 12 x 12 x 12 x 12 x12 . Add to this St. John's reference to 12,000 x 12,000 furlongs and we have a repeated emphasis on proportions of 12 by 12 which should not be ignored.

We do know that in the ancient world, particularly in Egypt, some designs were based upon a rectilinear grid to maintain the correct proportional balance, so perhaps what we are being directed towards here is a 12 x 12 base grid.

Once constructed, it is easy to see the grid's relevance as a background for the sacred geometry of the holy city and why a similar rectilinear grid, usually black and white, forms the basis for temple floors in both the Masonic and Templar traditions. It is actually a template upon which all the other geometric patterns can be easily constructed without recourse to modern aids like the protractor,

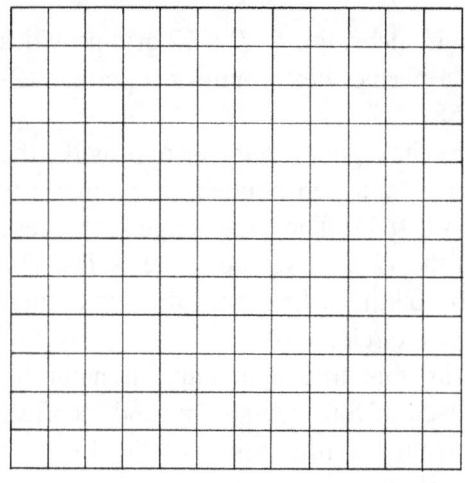

Illustration 156

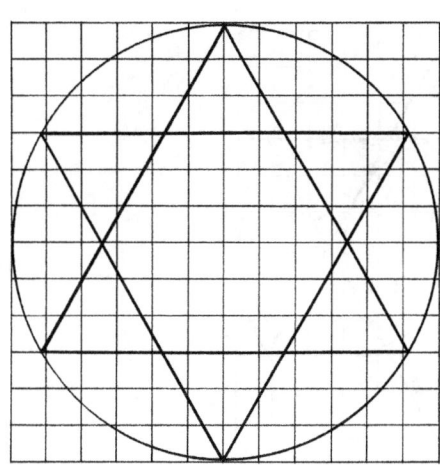

Illustration 157

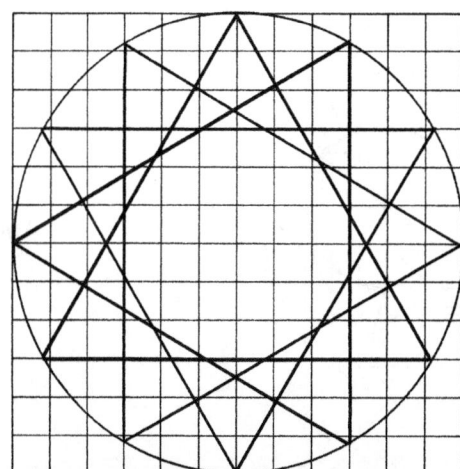

Illustration 158

which, of course, we not available to our ancestors. A 12 x 12 grid provides the necessary constructional lines for a perfect hexagram, or a perfect 12-point star (illustrations 156, 157 and 158).

The grid lines which define the hexagonal stars also provide the diameters of circles from which five and ten-pointed stars may be constructed (illustrations 159, 160 and 161). The arcs tangent to these circles are centred on the mid-point of the 12 x 12 square's outer sides. The precise tangent point may be found by taking a line from the same outer side mid-points through the centre of each circle.

When this pattern is repeated as four-fold symmetry, it naturally produces a twenty-pointed star, composed of four pentagrams and identical to London's. Alternatively, the lines from the mid-points of the 12 x 12 square's sides may be extended through the circles' centres and tangent points to define a 20-point star composed of two ten-point stars identical to London's Earthstars' decagon (illustration 162).

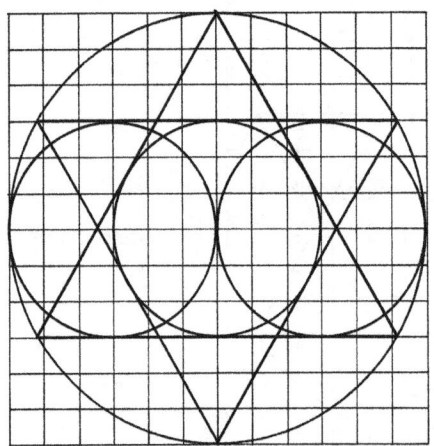

159

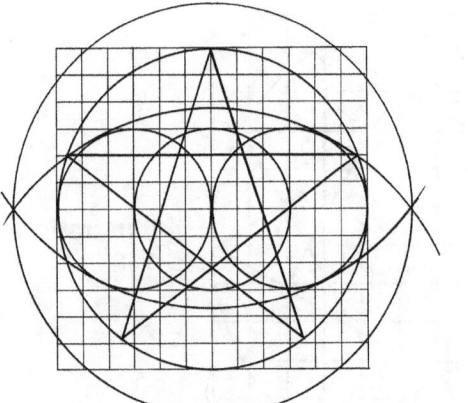

160

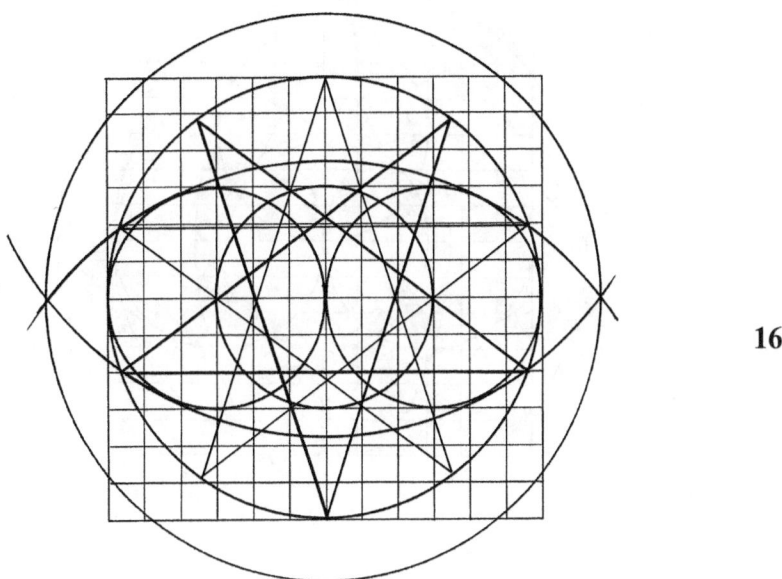

161

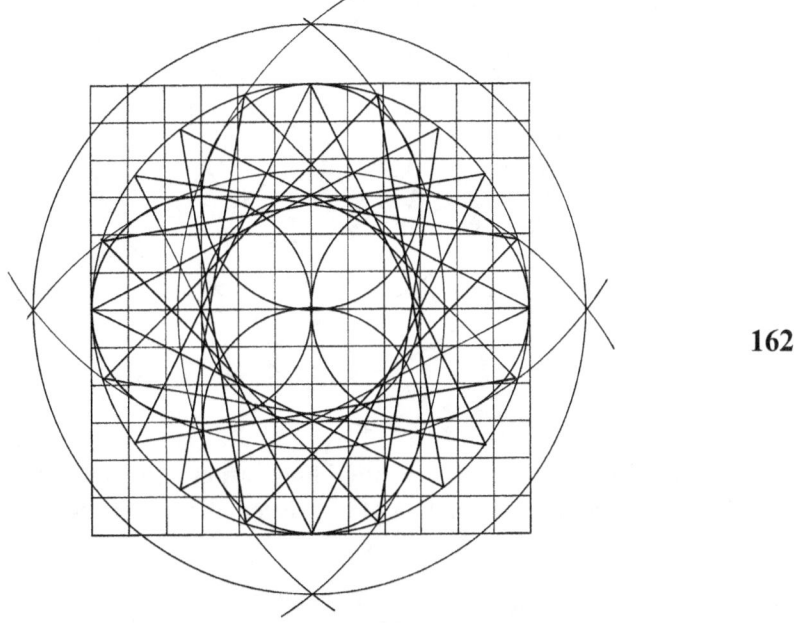

162

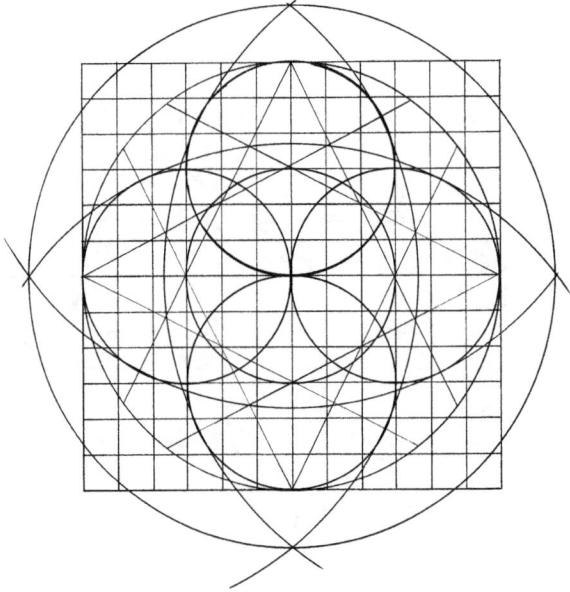

Illustration 163: Lines from the centre of the 12 x 12 square's outer sides, drawn through the centres of the two inner circles, extend to mark points of a 20-point star (see Illus. 162). Arcs from the same starting point define the outer circle of the squared circle diagram.

Illustration 164: The essential spiritual foundation underlying the squared circle would be this simple construction of interlaced circles; the geometry of creation

Remarkably, this is the constructional geometry of the squared circle, described previously in Chapter Seven (illustrations 72-76) so the 12 x 12 clues are definitely directing us to the same geometric figure.

One veiled reference to the squared circle design may have been coincidence. Three, in my opinion, constitute deliberately planted evidence. In one respect, it is not altogether surprising that the City of Revelation turns out to be founded upon the squared circle's geometry. As a mandala depicting the essential structure of the Holy City of God, or the Kingdom of Heaven upon Earth, the squared circle is the ideal template. Basically, it is the reconciliation of two seemingly incompatible elements; the circle and the square. In terms of the symbolism involved, though, that union becomes akin to an alchemical process.

The circle is traditionally considered a graphic representation of the eternal spiritual realms. In contrast, the square is finite and represents the tangible world of physical reality; the four corners of the Earth; the four elements; the four compass points; the four seasons; the four states of matter. The successful blending of these two opposites, by creating a square of equal area or circumference to a given circle, was thought to transmute the circle's intangible elements into finite form, thereby establishing an area of Earth created by Heaven and equal to it, in the process, constructing a bridge between the realms of spirit and earthly existence.

The complete figure of the squared circle represents a diagram of the energy connections between Heaven and Earth. I cannot emphasis this too strongly. The design symbolises the descent of spirit into matter via energy. It is a circuit diagram of the forces of Creation.

If this is the New Jerusalem, its descent from Heaven to Earth is not a literal allusion to its expected re-location. It is a permanent feature of a construction which is meant to span the gulf between the highest spiritual states and the material world.

Since this is also the basis of the ground plan defined by London's Earthstars sacred geometry, it could be assumed that this construction somehow represents a spiritual energy source connected to the natural energies of the planet. It must have existed here long before the city grew up around it. Only when sufficient sites had been marked by man would a pattern on this scale become identifiable.

London's circles are 16.2 miles and 20.625 miles in diameter respectively. These figures may superficially seem unremarkable, but there is nothing arbitrary about their proportional relationship. It is as precise and as unusual as the rest of London's geometry.

The ratio between the two diameters is almost exactly the same as the ratio between the two concentric circles in John Michell's examples of the New Jerusalem geometry which are 1: 1.272.

More important, the squared circle's constructional geometry, based on the four pentagrams of the twenty-point star, is identical to London's. The City of London and the City of Revelation are built upon the same geometric foundations.

By no means are these the only connections between London and the City of Revelation, either. In fact, the more detail you delve into, the more points of comparison are to be found between the two.

When the 12 x 12 grid is applied to London's Earthstars, it provides clear evidence of proportional harmony. Taking the 16.2 diameter as the full width of the grid, one twelfth is 1.35 miles. That is **10.8** furlongs, 432 poles, 2,376 yards, **7,128** feet, 85,536 inches.

108 is the lunar radius number; 432 is the solar radius number: 2376 is 3 times **792**, the Earth diameter number: **7128** is 9 times **792** the Earth diameter number, or 66 times **1080** the lunar diameter number and 16.5 times 432 the solar radius number. 85,536 is **108** times **792,** or 99 times **864** the solar diameter number.

Once again, all the key numbers of universal proportion are repeated. 12 x 12 is obviously an immensely important number in the general scheme of things, too. 12 x 12 is 144 and 144 is present as a multiple of all the key numbers of the Sun, Moon and Earth.

864,000, the sun's diameter, is precisely 6,000 times 144. 7920, the Earth diameter is precisely 55 times 144. 2160, the moon's diameter is precisely 15 times 144.

In the Earthstars dimensions, 144 crops up repeatedly. The inner circle's diameter in poles, 5184, is 144 x 36. The outer circle's circumference in furlongs, 1036.8, is 144 x 7.2. In yards it's 228,096, which is 144 x 1,584. In poles, it's 41,472 which is 144 x 288 or 144 x 144 x 2.

144 is a foundational number, not just of the biblical City of Revelation, but of cosmic proportion. As John Michell tells us in the foreword to William Stirling's THE CANON;

"The figures of the New Jerusalem, Holy Oblaton and other temples, real and imaginary, reveal the magnitudes of the sun, moon and other planets together with their distances of their orbits."

What we are being given in the Book of Revelation is not merely the proportions of the Holy City, but coded information from an ancient scientific system, almost certainly pre-dating Christianity, which understood not just the structure of the universe, but also possessed the mathematical keys to the forces underlying it.

Surprisingly, our ancient British measures of furlongs, poles, yards, . feet and inches appear to be compatible with this system, if not directly based upon it. John Michell's work consistently stresses the importance of British units of measure, whilst Henry Lincoln and David Wood both remark on it in their study of Rennes Le Chateau's geometry In THE HOLY PLACE, Chapter 9, Lincoln states;

> **" For the construction of The Holy Place of Rennes Le Chateau, the mile and its subdivision, the pole, were the measurements used."**

THE GATES OF THE CITY

The gates of the Holy City are given particular prominence in St. John's Revelations and clearly form an important part of its construction. Do they have some esoteric relevance to London, too?

Amazingly, they do.

A close look at the references to the gates of the city reveals that important parts of London's Earthstars geometry fit the biblical descriptions perfectly.

> **"And (the city) had twelve gates, and at the gates twelve angels, and twelve names written thereon which are the names of the twelve tribes of the children of Israel."**

Revelation 21; vs 12.

> **"On the east three gates; on the north three gates; on the south three gates; and on the west three gates."**

Revelation 21; vs 13.

It does not take a genius to work out from these passages that the gates of the city are somehow arranged in four groups of three. The most obvious way to arrange three points is as a triangle, so the complete picture of the new Jerusalem's gates, re-constructed from these references, could be interpreted four triangles, one at each of the four compass points. London's Earthstar patterns have four triangles, one at each of the compass points.

In the North, the Barnet Triangle. In the South, the Croydon Triangle. In the East, the East Ham Triangle; and in the West, the Hanwell Triangle. Could London's four triangles somehow act as the gates to The Holy City?

THE PEARL OF HEAVEN

A further description of the gates in verse 21 adds more detail with direct relevance to London.

> **"And the twelve gates were twelve pearls; every several gate was of one pearl."**
>
> Revelations 21; vs 21.

Every several gate is a clumsy phrase, but it does indicate once more that the gates are somehow grouped, several together, in case we had missed the earlier reference to them as four units of three.

There's another clue in the reference to twelve pearls.

In his New Jerusalem diagram, John Michell depicts the twelve gates as lunar spheres. This makes good sense given that pearls have esoteric lunar associations and these 12 may be symbols of the 12 months which derive from a lunar cycle. Also, the new Jerusalem's concentric circles embody lunar proportions and John locates the 12 pearls on the appropriate lunar circles.

Nevertheless, something about this explanation left me feeling it wasn't quite right. I couldn't explain why and I didn't have a better solution until, one day, I stumbled upon the fact that the Pearl of Heaven is a phrase often used in reference to the Virgin Mary.

This information gave me an insight into an equally reasonable explanation of the 12 pearls that fits London's triangles perfectly.

Every single one of the churches defining the four triangles are dedicated to the Virgin Mary, the Pearl of Heaven.

For those with an eye for detail, St. Margaret's which stands next to the ruins of Barking abbey is not an exception.

Barking Abbey pre-dates St. Margaret's by several centuries and its original dedication was to the Virgin Mary.

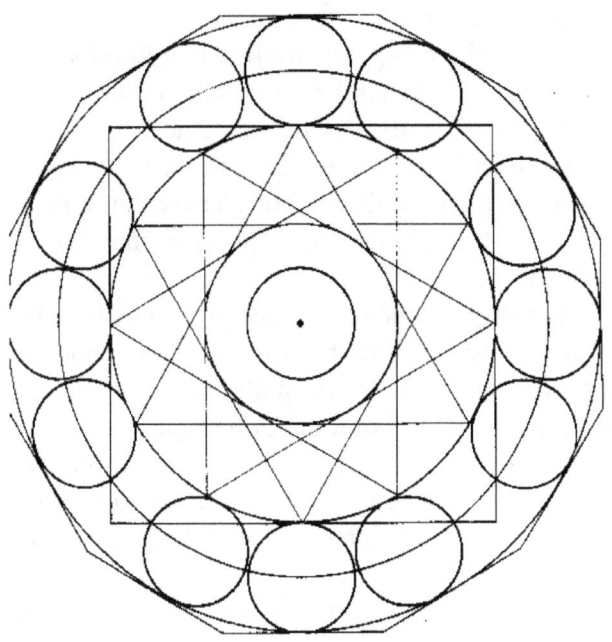

Illustration 165: John Michell's interpretation of the New Jerusalem geometry incorporating the twelve gates as twelve lunar spheres on circles proportionate to the lunar diameter (reproduced with permission).

Each of the four triangular gates in London conform to a pattern;

All four have two churches dedicated to the Virgin. All four have a third site which probably pre-dates Christianity in these islands and is not currently built upon. In the Barnet Triangle, it is the Camlet Moat location. In the East Ham Triangle it is the "healing well" now lost beneath Central Park Road. In The Hanwell triangle, it is the mark stone of indeterminate age which was last noted at the corner of Grosvenor Road and Uxbridge Road. In the Croydon Triangle, it is the impressive hill-top henge at Pollard's Hill.

It has to be said that even the sites presently marked by churches are mostly of sufficient antiquity to imply that their origins are probably pre-Christian as well. So, despite the Biblical references, we are not dealing here with an exclusively Christian mystery, but something far older, a mystery stemming from humanity's earliest spiritual experiences and which may prove common to the many different religions which acknowledge the natural sanctity of certain places.

Meanwhile, back at the city gates, we find that London's four triangles don't simply conform to the general description of the New Jerusalem's gates. In one sense, they actually function as gates, too. Not the kind of gates that let visitors in and out, you understand. We are dealing here with sacred geometry, the circuitry of divine force. These could be some kind of energy gates. Each triangle could be a three pin plug into the cosmic mains; The ultimate power source.

In sacred geometry, remember, triangles are the first physical manifestation of divine force. Think back to how these Earthstars patterns unfolded. London's landscape geometry in all its entirety and intricacy was generated by the simple geometry of the four triangles.

Their geometry dictates the geometry of the whole design, of the entire city. If these patterns represent the forces of creation, the geometry of the four triangles determines the precise forms through which they flow into the capital.

In this respect, by far the most important is the northern gate, the Barnet Triangle. It generates and controls the energy patterns of the ten-point star, the eight-point star, the Camelot pentagram, and the eighteen-sided polyhedron.

The Croydon Triangle generates a pentagram and also shares the influence of the eighteen-sided figure.

The East and West triangles each generate a pentagram and share the influence of the fifteen and thirty-point stars.

Not only do the triangles exercise direct control over the specific geometric energy patterns of the city, they also dictate how those patterns and energies interact.

The precision with which they accomplish this is nothing less than miraculous. London's Earthstars mandala is made up of nineteen separate geometric devices, every single one linked in some way to the others and built upon the foundation of the squared circle. The overall design is a construction of such complexity and precision that I challenge anyone to fit those figures together more coherently or beautifully.

The idea that a triangle of ancient sites can somehow act as a gateway or focus for some kind of spiritual energy is going to be hard for most people to accept. Yet once again, the landscape itself provides further indications of this precise notion; at the main gate of the city, the Barnet Triangle.

In sacred geometry, symbolic form is frequently synonymous with the nature and function of the energies involved and the Barnet Triangle is a very specific form; a near perfect equilateral triangle.

In Christian terms, it is a symbol of the Holy Trinity or Holy Spirit.

If this is a gate, one interpretation of its significance would be that the energy entering the city through it ought to be the power of the Holy Spirit. Pretty far-fetched wouldn't you agree?

But what if there was a second symbol of the Holy Spirit emblazoned upon the landscape in the same area? Actually, there is. It is the outline of a dove, sculptured in the natural features of the area.

The head is most clearly defined, in part by the main road from High Barnet to East Barnet. Other parts of the figure, I admit, are open to debate. In principle, it is similar to the kind of landscape figures featured in Katherine Malt wood's Glastonbury Zodiac, although its outline is slightly more recognisable than some of the Zodiac's figures. Also, there is an assumption with the Glastonbury Zodiac that these designs were laid out in antiquity and have decayed over the centuries.

I do not necessarily think that this dove falls into that category. I believe it was formed naturally, by the combined effects of nature, mankind and unseen forces which work behind the scenes influencing both. In a way, you could say that the spirit of a place is able to express itself both through nature and through the spirit of the people in a particular area, who may not be aware of their sources of inspiration or that their constructions form part of some design beyond their immediate comprehension.

The figure of this dove, I believe, has evolved over a long period as an expression of the spirit of place active in this area. The dove appears to be flying into the Barnet Triangle, and the implications of the power of the Holy Spirit entering the soul of the city, or spirit of a nation, are in keeping with the appearance of the New Jerusalem which is said to herald the foundation of the Kingdom of Heaven upon Earth. These, of course, are Christian concepts. I personally believe they incorporate a much more ancient mystery tradition from which aspects of Christianity may itself have evolved.

It has always puzzled me why there was no mother in the Holy Trinity. We have the Father, Son and Holy Ghost. Surely it should be Father, Son and Mother, but instead of the Mother we have the mysterious Holy Ghost. Why?

Is it because the Holy Ghost or Holy Spirit represents the Universal Life-Force and that, prior to Christianity, its strongest personification was as the Earth Spirit, an aspect of the Mother Goddess.

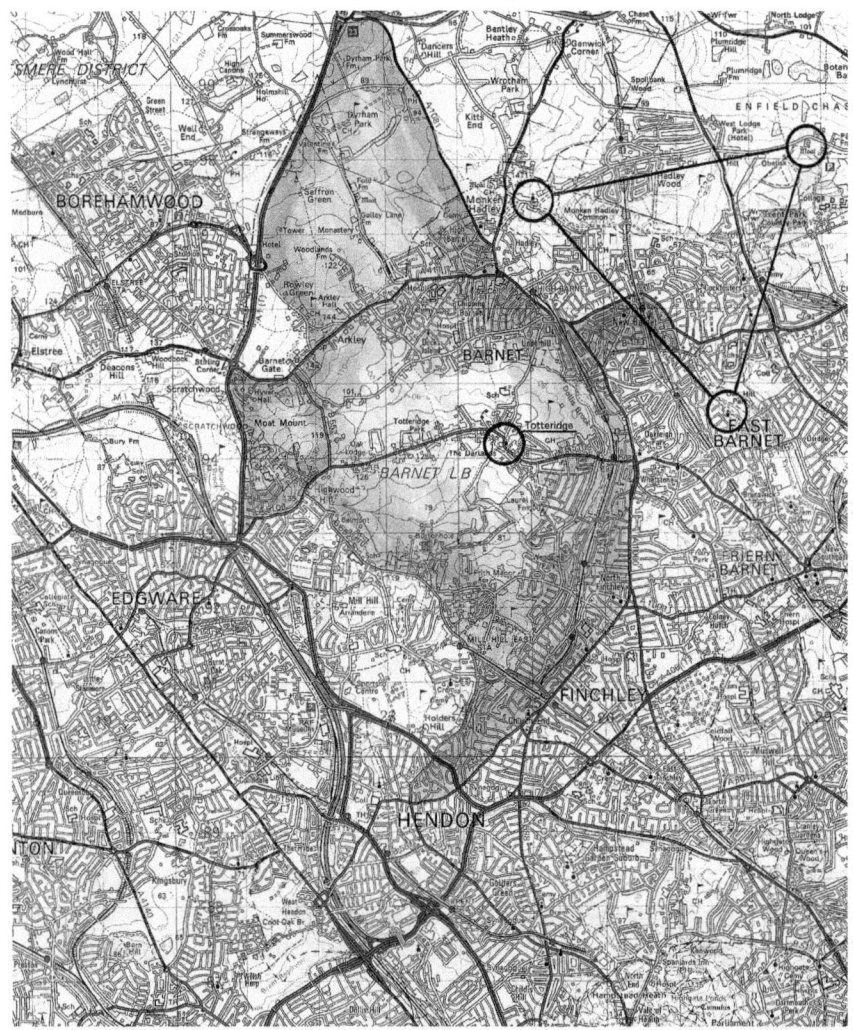

Illustration 166: Landscape features resembling the shape of a dove in the area of the Barnet equilateral triangle. Two symbols of the Holy Spirit concealed within the landscape. O.S. map © Crown Copyright. All rights reserved. License No. 100029558.

Certainly, William Stirling in THE CANON believes this to be the case. On Page 216 he refers to; **"Bride, the third person of the triad."** And on page 287; **" The Great Mother Earth, or the heavenly spouse of the Logos, called The Holy Ghost."**

If the mother goddess was associated the old religion which early Christianity was attempting to suppress and supersede, it is predictable that she became the invisible one, the Holy Spirit, the church's great secret. Nor is it surprising that her cults became occult, hidden, outlawed by the church, but at the same time concealed and incorporated within it as part of the Marian elements, which in terms of the psychology of the collective unconscious, provide all the necessary goddess archetypes in Christian form.

The power of the Holy Spirit is supposed to be the creator's divine energy. It is not a quantum leap to think of it as the Universal Life-Force. After all, many Christians I know personally, claim they actually feel the Holy Spirit in similar ways to which others, myself included, feel the energies at sacred sites, or in leys.

Is this the secret behind the mystery of the Holy Ghost ? Is it a secret merely because of its pre-Christian associations with the Earth Spirit and Mother Goddess ?

This would certainly help to explain aspects of the Earthstars phenomena, particularly the connections between apparitions of the White Goddess, Bride, Guinevere and the intrinsic energies of some sacred places.

They are different ways of anthropomorphising the energy of the Earth Spirit.

The apparitions are the energy. The geometry is the energy, too.

Every line of enquiry seeking an explanation for the Earthstars' discovery leads straight back to the same source.

What our ancestors were working with and what we are dealing with here is a science of the Universal Life Force and its local manifestation as the Earth Spirit. For those who might think this is a pagan concept, I have to say that in a holistic universe it cannot be anything other than an aspect of the omnipresent creator, God.

In support of this line of thinking, it seems likely that the mysteries of the Earth Spirit and Universal Life Force have been incorporated into Christianity as the mysteries of the Holy Spirit.

Interestingly, bearing in mind the connections to Camelot and Gwenevere, they also appear to have been concealed within the Arthurian tradition as the mysteries of the Holy Grail.

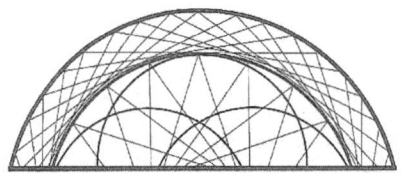

Chapter Fourteen
The Bride

It is a curious fact that throughout the Book of Revelation, the New Jerusalem is referred to as female and more specifically, as "The bride of the lamb,"

> "Come hither, I will shew thee the bride, the lamb's wife.
> And he carried me away in the spirit to a great and high
> mountain and shewed me that great city, the holy Jerusalem,
> descending out of heaven from God."
>
> Revelation 21; vs 9-10

> "And I John saw the holy city, new Jerusalem, coming down from
> God out of heaven, prepared as a bride adorned for her
> husband."
>
> Revelation 21; vs 2.

There is an unmistakable reference to a marriage and in the Earthstars geometry, as we have seen, there is very definitely a symbolical marriage of opposites, both in the squared circle design and in the thirty-pointed star which marries pentagonal and hexagonal symmetry. The identity of the groom is hinted at by "the lamb."

> "Let us be glad and rejoice, and give honour to him; for the
> marriage of the lamb is come, and his wife hath made herself
> ready for him."
>
> Revelation 19; vs 7.

The lamb of God is a term used to describe Christ. But who exactly is the bride, the lamb's wife, and what is the meaning behind this marriage? Several authors in recent years have made the suggestion, based upon extensive research by historians and biblical scholars, that Christ may actually have been married and that his wife was Mary Magdalene. For anyone wishing to enlighten themselves on this matter in detail, I would recommend THE HOLY BLOOD AND THE HOLY GRAIL, by Michael Baigent, Richard Leigh and Henry Lincoln, THE TEMPLAR REVELATIONS by Clive Prince and Lynn Picknett, THE SECOND MESSIAH by Christopher Knight and Robert Lomas, And BLOODLINE OF THE HOLY GRAIL by Laurence Gardner.

What intrigued me more is that there is an ancient British Goddess actually known as Bride and, according to Robert Graves, Bride was one of the names of the White Goddess. Apparitions of the White Goddess played a key role in initiating the Earthstars discovery, so she is directly connected to it.

So what do we know about Bride? According to Kathy Jones, author of THE ANCIENT BRITISH GODDESS, Bride was actually a deity with many names, appropriate to different locations. In Ireland she was known as Bride of the White Hills, or Bride of the Golden Hair. In Scotland, she was Bride of the Fair Palms or Mary of the Gael, often equated with the Virgin Mary. She was also Brig, Brigid or Brigit. She is associated with the colour white and with the white swan. Her patronage is as the guardian of women in childbirth and the divine midwife which, may account for some of her associations with Mary.

The latinised form of Bride or Brigit is Brigantia and the Hebrides aren't the only islands named after her. As Brittania, she gave her name to the British Isles and is the guardian spirit of them.

In this sense then, she is the genius loci of these islands; an interesting association which clearly identifies her as an aspect of the Earth Spirit and one of the many faces of Mother Nature or the ancient Mother Goddess, in this part of the world at least.

Her curious association with the Virgin Mary, suggests that with the coming of Christianity, Bride, in her virginal aspect, was equated with Mary the Virgin who was also Mary the divine Mother. As a result, many of Bride's places of veneration may well have been re-dedicated to St. Mary. Bride clearly had more than one facet to her character. As well as having aspects as both virgin and mother, she also seems to have been a fire goddess associated with the sun. At her principal shrine in Kildare, an eternal flame burned. So in her triple form, she represented the Sun, Moon and Earth, although it is possible that her differing aspects were also known by other names.

Despite the fact that most of the general populace have forgotten her completely as a goddess, the strength of her original following in these isles was such that her name has lived on, not just as Brittania and place names like Brighton (Brig's town), Brixham, Bristol and Brixton, but in every woman who ever weds. Every woman who walks down the aisle and stands before the altar is known by the name of this ancient British Goddess, Bride.

It can be no coincidence either, that the bride traditionally wears white. It is the colour associated with Bride and the reason she is also known as the White Goddess.

All this suggests that today's wedding ceremony owes some of its origins to a pre-Christian ceremony involving a symbolic marriage to the goddess of the land, who is partially remembered in these isles as Brig, Brigantia, Brittania or Bride, but who elsewhere may have represented the Earth Spirit or the Great Mother.

The relevance of this to London's Earthstars is that there is good reason to believe this feminine archetype, in whatever form you wish to imagine her, is embodied in certain aspects of the sacred geometry, in particular the pentagonal patterns.

If the sacred geometry of a cathedral represents the soul of the structure then, by the same token, similar geometry in the landscape represents the soul of the land and an aspect of the Earth spirit.

The main pentagonal patterns, four pentagrams directed to the four quarters, are on the Circle of the Earth Spirit: Traditionally, the Earth Spirit is feminine. It is personified as Mother Nature, or Mother Earth and in days gone by as the Great Mother or the Earth Goddess.

It is probably therefore, no coincidence that the four triangular gates which generate and feed into this pattern are all predominantly defined by sites dedicated to St. Mary, the more acceptable Christian version of the Holy Mother. All but one of these churches are sites of exceptional antiquity, in use as places of Christian worship for at least eight hundred years or more and since it was general practice for the earliest churches to be built upon existing places of worship, they were probably considered sacred long before Christianity.

A prime example of this is St. Bride's in Fleet Street where excavations in the crypt, following a WW2 bomb, revealed Roman, Saxon and Norman structures, all built in succession, close to an original shrine to the ancient British White Goddess.

It was also common practice for the church to dedicate such churches to a Christianised variant of the original deity. Thus Brig or Bride, became Saint Brigit or Saint Bride, or since she was also known as Mary of the Gael, she may also have given rise to some dedications to St. Mary. It is generally accepted amongst researchers in this field that St. Mary's churches of great antiquity probably originated as places sacred to a pre-

- Christian, mother goddess. The church of Notre Dame (Our Lady) in Paris is another excellent example, having been built upon a driudic shrine to the Mother of Heaven.

To assume the St. Mary's churches at the four Earthstars triangles were once shrines where an Earth Mother deity was revered is not unreasonable. Symbolically, they connect to the Circle of the Earth Spirit where her pentagonal star field and mandala are epitomised in the twenty-point star's four pentagrams which channel the circuitry of the Earth Spirit into the capital.

Further evidence that the pentagram is a symbol of the feminine polarity of the Universal Life Force can be found in the Jewish Kabbalistic tradition. William Wynn Westcott, in THE MAGICAL MASON, states that the pentalpha, in Kabbalistic symbolism, relates to a female polarity of the divine which may be either;

" **The Mother idea**" or **"the bride of God, the church, the kingdom."**

This of course, combines Bride, the White Goddess, with the Mother Goddess, the church and the land and supports the idea that they are all in fact associated as different facets of the same principle. The various aspects of the Earthstars' discovery re-inforce this idea.

The meaning behind all this is not easy to grasp, but it seems to be telling us again that while the Earthstars geometry may represent the life force in general, certain patterns relate specifically to the Earth Spirit, who was personified and seen as a goddess by our distant ancestors and who may have been incorporated into mainstream Christianity under the auspices of the Virgin and divine Mother, Mary.

This understanding might go some way to explaining the apparitions I saw at certain sites which led me to the Earthstars discovery. But what on Earth does the bride's marriage to the lamb mean ? Again, the Earthstar's geometry holds a clue.

The overall symbolism of the squared circle is a symbolic union of opposites; of heaven and Earth. So these references could, once again, be pointers towards the squared circle sacred geometry as a union of the Earth's intrinsic energies with those of the universe at large.

There is, however, another "marriage" of opposites within the Earthstars' geometry which confirms the validity of this perspective. One of the most important parts of the Earthstars' symmetry is the way in which the thirty-pointed star combines the pentagram and hexagram. In quite different ways a simple combination of pentagram and hexagram features in John Michell's sacred geometry of Jerusalem, as well as David Wood's and Henry Lincoln's geometry around Rennes Le Chateau.

This union is of immense significance esoterically.

In GENESET (Bellevue Books, 1994), David Wood's follow-up to GENISIS, he refers to the combination of a single pentagram and hexagram which he had discovered in the Rennes Le chateau region:

"Such a figure had never been revealed in any books on sacred or ordinary geometry until it was published in GENISIS. It would, however, have had devastating significance to the esoteric world, for the star union, the blending of both life forces, the Yin and the Yang of the Chinese school, was every alchemist's dream. Even the notorious Eliphas Levi (Alphonse Louis Constant), mysteriously stated that "whoever discovered the secret of uniting the pentagram and the hexagram was halfway to solving the mysteries."

As David Wood rightly says, it represents a blending of polarities, of yin and yang, of negative and positive, of male and female. It is an alchemical marriage and an act of pro-creation. If, in terms of esoteric energies, the pentagram and hexagram represent two polarisations of force, the Earthstars' patterns have already helped us deduce which is the feminine polarity.

As we have seen, the pentagonal patterns have a definite association with the feminine aspect of divinity in the form of the Earth Spirit. The six-point star, which has been known as the seal of Soloman and the star of David, in this instance, may therefore represent the male polarity. In THE BOOK OF DRUIDRY, Ross Nichols has this to say in reference to the use of 30 sarsens at Stonehenge;

" 30 is the Avebury number, 5, probably representing energy on the definite, material plane, multiplied by balance, 6 and 6 is the number of the sun as balance."

These correspondences echo the numerology revealed by dividing the New Jerusalem's 12 x 12 into solar and terrestrial numbers.

144 divided into the solar 864 produces 6.
144 divided into the terrestrial 7920 produces 55.

In this context, 6 has solar associations while 5 is linked to the Earth. The Earthstars' thirty-point star combines pentagonal and hexagonal symmetry in what might be the most perfect way possible.

A thirty-point star can be drawn as five hexagrams, or six pentagrams. Its essence therefore, is a combination of both, and as such, it is the ultimate marriage of pentagonal and hexagonal symmetry. Moreover, it is "earthed" to the four pentagonal patterns on the Circle of the Earth Spirit by the fact that both share five fundamental axes in common; those of the initial pentagram and the ten-point star of which it is a part.

If this is a graphic symbol of a marriage, it represents a blending of two or more energies. What does this union generate? A higher energy? The powers of creation in general, or the evolution of new life on Earth ?

In terms of the biblical references, it is a union between a Christian saviour and his bride, the mysterious Earth Spirit, the soul of the Earth, of which we are all a small part.

On an archetypal, mythic level, it is a the creation myth of the Earth goddess being fertilised by the solar hero, the sun god, which formed the basis of many ancient religious beliefs and fertility rites, like those of Egyptian Isis and Osiris, or the May King and Queen.

In the New Jerusalem Earthstars' geometry, it is being presented in yet another way, as a geometric plan, showing the interaction of the different geometric forms which represent a possible blending of the Earth's intrinsic energies with those of the sun, moon, other planetary bodies and "the heavens" as a whole.

The same theme is continued in the dimensions of the Earthstars patterns as the proportional numbers of the main planetary bodies, all of which also crop up again in the 30-point star.

The distance between each point, for instance, is 12 furlongs or 7,920ft. The perimeter is 7,920 x 30 ft or 79,200yds; all Earth dimensions. The solar and lunar numbers can also be found in the star's various component parts.

It is an inescapable fact that a complex interaction between the cycles of the Sun, the Earth and the Moon, generates all life on this planet, so it is startling to find, in the Earthstars patterns and in particular the thirty point star, a graphic symbolism of those myths in an overall design representing the life force of the planet and universe.

Once more, we find the Earthstars geometry pointing us towards a new perspective on those forces and our role in relation to them.

The Earthstars mandala is a blueprint which incorporates the proportions of structure in the cosmos, the Universal Life Force, the Earth

Spirit and, since our individual spiritual dimensions are part of both, the human collective unconscious.

The New Jerusalem is a symbol of the perfect state of being, a representation of our spiritual dimensions. Its structure lays bare a divine plan, the interplay of celestial forces which create and compose the entire universe, the inner structure that underlies matter, linking all invisibly as one.

As such, it is not a prophecy to be fulfilled.

It has been here all along, as an eternal link between the worlds of matter and spirit.

The correct moment for it to emerge into the sphere of human awareness seems to have arrived.

Every mystical tradition tells us that all things are connected. The New Jerusalem geometry gives you a good idea of the complexity and beauty of what connects them.

The energies that comprise this structure are the very essence from which all life springs. These forces are the diverse manifestation of the one single impulse at the source of creation and, as such, they have existed since the beginning of time.

Their essence may be pure energy, but they are alive, although not in quite the same way you and I are.

They were personified by our ancestors as gods and goddesses, or hierarchies of angels. They are the archetypes that populate mythology and folk-lore as well as the dimensions of our collective unconscious, in the psychological sense. They are a universal life-support system without which we would simply cease to exist.

In a manner of speaking, the Earthstars patterns are a map of our spiritual dimensions, the Kingdom of God within, within the Earth, within you, within me, within everything.

The energies they represent are part of us and we are part of them. Inseparably. They give us life and we, in turn, are their physical expression and conscious awareness in the material world.

The appearance of the Holy City is said to herald the foundation of the Kingdom of Heaven upon Earth.

If the Earthstars pattern represents the New Jerusalem and the inner spiritual dimensions of the planet and its populace, it does not <u>herald the foundation</u> of the Kingdom of Heaven on Earth. From this perspective, <u>it is the foundation</u> of the Kingdom of Heaven on Earth.

As such, its geometry, structure and symbolism are immensely important to us. Whether London is literally the New Jerusalem or not, it does exhibit the New Jerusalem's spectacular sacred geometry in the relationships shared by some of its most ancient places of worship and is clearly a very special place.

In the light of these revelations, it is hardly surprising that London has evolved into a place of temporal power.

It now warrants consideration as a place of considerable spiritual power.

THE LION AND UNICORN

Implicit in these matters is also the understanding that the city may be under some kind of divine protection.

Astonishingly, there are two further figures emblazoned on the landscape around London which support this idea. They came to my attention one day while I was idly staring at one of my Ordnance Survey maps. I wasn't looking for anything in particular, but suddenly noticed that some major roads in Essex resembled the outline of a lion's head, and a rather sphinx-like one at that. It is facing due east to where the sun would rise at the equinoxes just like the enigmatic sphinx of Giza.

The nose, mouth and general facial area is marked by the B175 from Passingford Bridge to Chase Cross. The chin by the B174. The front of the body and paw by the A1112 as far as the A13 junction near Rainham. The mane and back of the head is marked by the A113. The back, by parts of the A406 and A503. The head and front of the body lie outside the Earthstars' circles. The bulk of the body lies within them and is actually much more difficult to discern in such heavily built-up areas. The general shape, though, is definitely recognisable and much more distinct than some of the figures claimed for landscape zodiacs.

The lion, of course, is one of Britain's heraldic totems and, along with the unicorn, features on the Royal coat of arms. Once I realised this, I wondered if there might be any trace of a unicorn on the map, too. After examining my large scale 1;50,000 Ordance Survey maps, I found that the rough outline of a horse's head could be discerned in the path of the river Cole and other nearby features to the north west of London. Remarkably, some distance from the forehead was a spot called Horn Hill with footpaths and boundary lines defining a short horn-shape leading to it. At the tip of the horn is a small, but extremely pleasant chapel.

Again, various roads, paths, boundary lines and other landscape features have been used to mark the limits of this animal's outline.

Where the figure merges into the urban sprawl of Greater London, its definition becomes indistinct. Nevertheless, I believe that, like the Barnet Triangle's dove, the lion and unicorn have evolved naturally over a long period, rather than having been created in antiquity and obscured by time.

They are not perfect images, I admit. They do require a certain degree of imagination to appreciate their existence. But even vaguely recognisable representations of Britains heraldic beasts, standing as guardians on either side of the capital are a remarkable enigma to have been found.

Doubly so, considering that they are part of the Earthstars discovery and add considerably to its significance.

They suggest, as do the Earthstars patterns generally, that London is a place under some kind of divine protection, although I appreciate this may be extremely difficult to believe if you happen to live in any of the capital's worst run-down areas.

Coincidence or chance seem the least credible explanation for these figures or for the Earthstars patterns.

The accumulated evidence indicates a very real phenomenon here, albeit one that is based as much in our elusive spiritual dimensions as the physical world.

Illustrations 167 and 168 (pages 253 and 254): Figures of a lion and a unicorn, London's two heraldic guardians, appear not only on the Royal coat of arms, but emblazoned on the landscape. The lion on the Essex landscape faces due east like Egypt's sphinx. The unicorn faces west.

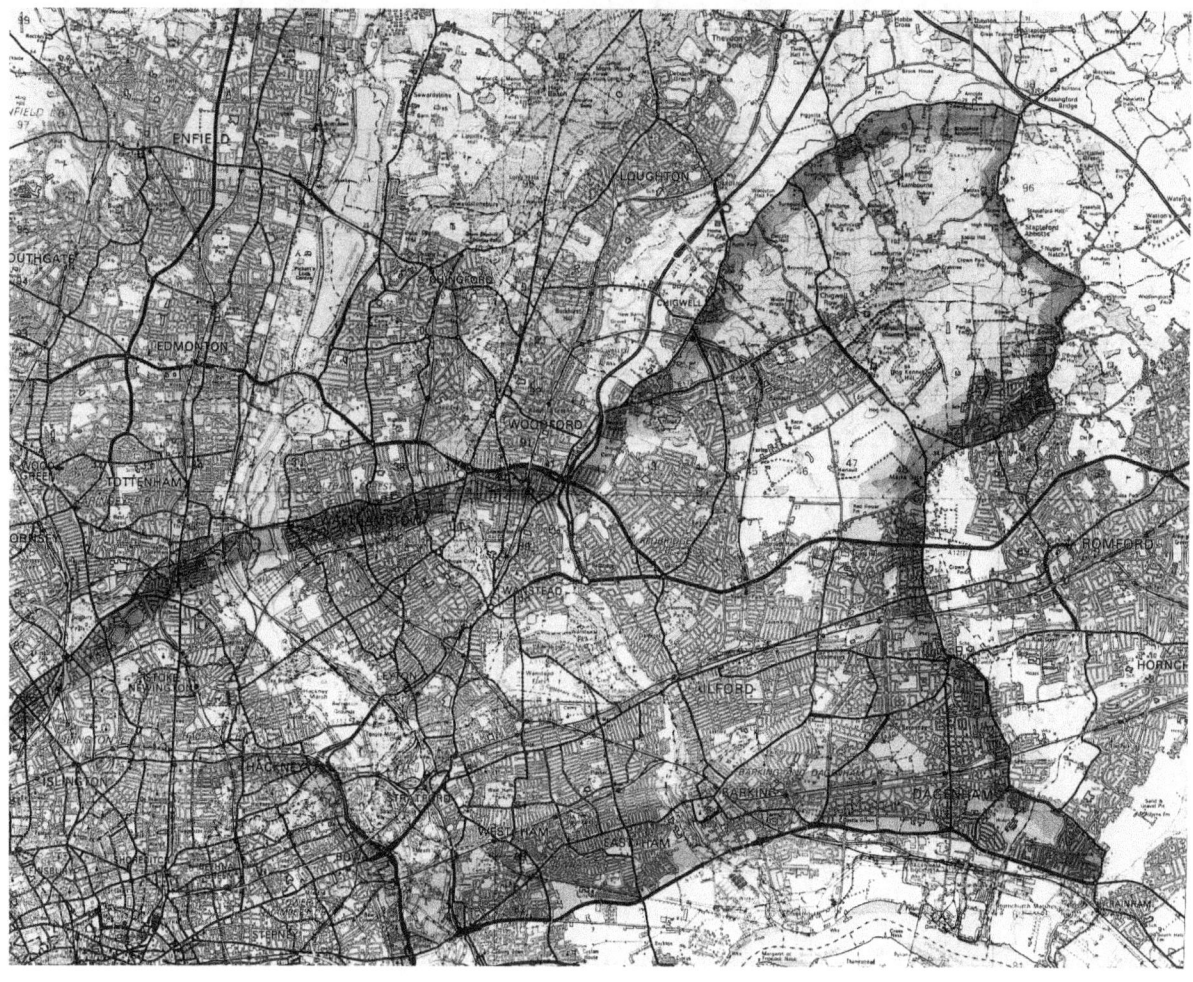

Illustration 167: The outline of a sphinx-like lion on the Essex landscape to the N.E. of London. O.S. map © Crown Copyright. All rights reserved. License No. 100029558.

Illustration 168: The outline of a Unicorn defined by various landscape features on the landscape to the N.W. of London and facing west. The point of its horn is actually called Horn Hill. O.S. map © Crown Copyright. All rights reserved. License No. 100029558.

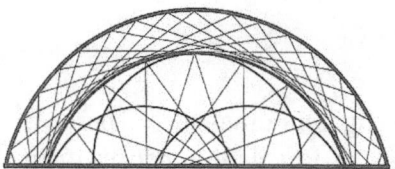

Chapter Fifteen
The End of An Age ?

Whatever you think of London, its spiritual foundations, in terms of the Earthstars geometry, certainly seem applicable to St. John's measures and proportions of the Holy City of Revelation. So what are the implications of this ?

Well, as the old joke goes, there's some good news and some bad news. The good news is that the appearance of the New Jerusalem is said to herald the coming of the kingdom of Heaven on Earth. The bad news is that it entails the end of the world as we know it.

> " And I saw a new heaven and a new earth; for the
> first heaven and the first earth were passed away;
> and there was no more sea."
>
> Revelation 21; vs 1.

The first section of the book of Revelation dwells heavily on scenes of apocalyptic destruction which could include volcanic eruptions, earthquakes, and even a possible deadly impact from an asteroid or comet.

> "and as it were a great mountain, burning with
> fire was cast into the sea;"
>
> Revelation 8 ;vs 8.

> "and there fell a great star from heaven,
> burning as it were a lamp "
>
> Revelation 8; vs 10.

> "lo, there was a great earthquake; and the sun
> became black as sack cloth of hair and the moon
> became as blood "
>
> Revelation 6; vs 12.

London   City of
Revelation

These are the lighter moments. There is so much death and destruction in the Book of Revelation that it's difficult to work out what to start worrying about first. If the earthquakes, comets or volcanoes don't get you, the plagues, famines, the four horsemen or the war to end all wars might.

Understandably, as we pass the turning point of the millennium in our calendar, the prophets of doom are having a field day and some people are becoming more than a little apprehensive. Is there any reason for us to succumb to this atmosphere of overwhelming dread ? Actually, there is. But these days, it tends to be environmentalists that warn us the end is nigh, rather than visionaries like St. John and personally, I find that even more worrying.

We have a hole in the ozone layer. We have global warming, melting ice caps and rising sea levels. We have widespread pollution afflicting the land, rivers, seas and atmosphere with all manner of human refuse from poisonous pesticides and genetically engineered crops, to nuclear waste. The rain forests which produce much of our planet's precious oxygen are being destroyed at a ridiculous rate. We've hunted many species to the point of extinction and are responsible for thousands of others disappearing from the face of the Earth.

Some sea creatures around the British coastline have mutated and changed sex due to hormonal imbalances introduced by pollutants which have already found their way into our food chain.

Add to all that, the fact that several countries have stockpiled enough nuclear weaponry to turn this planet into another asteroid belt and its hardly surprising we might soon be topping the endangered species list ourselves.

This is without considering the huge social problems facing many countries; even the developed western nations have overwhelming problems with drugs, crime, overpopulation and unemployment. The questionable ethics of a commercial system whose single meaningless aim is the pursuit of ever increasing profits doesn't help, either.

To any sane human being, it is all too obvious we are in an unsustainable culture which is busily sowing the seeds of its own demise. Is it any wonder that the threat of an impending crisis looms large in people's expectations ?

THE MAYAN PROPHECIES

Evidence that the end may indeed be nigh can be found in the records of the Mayans, a race who mysteriously disappeared from the face of the Earth, so perhaps are qualified to know about these things.

Study of their calendar and beliefs reveals that global cataclysms were thought to have brought four previous ages to an end and constitute a recurring cosmic cycle of events. By Mayan calculations, we are currently living in the fifth age which is due to terminate on December 23rd, 2012.

This information was first revealed by the visionary author Jose Arguelles in his book THE MAYAN FACTOR, but has since found greater readership in the more recent MAYAN PROPHECIES and was also included in FINGERPRINTS OF THE GODS by Graham Hancock, a book which deals with the subject of global cataclysms in great detail.

The fundamental premise of the book is that an advanced civilisation may have existed in our distant past and was totally annihilated by a natural catastrophe leaving only vague clues to their existence. Hancock's arguments and evidence are extremely convincing and his extensive research unearthed some interesting astrological details of relevance to the Mayan's calculation of the end of our age. On page 246, we find this snippet of information;

"Astrologers who have charted the Mayan date for the end of the fifth sun calculate that there will be a most peculiar arrangement of planets at that time, indeed an arrangement so peculiar that it can only occur once in 45,200 years. From this extraordinary pattern we may expect an extraordinary effect. "

Few people actually take astrological influences very seriously. By contrast, no-one questions the ability of the moon to tug the Earth's ocean's so powerfully it creates the tides and scientists are now discovering that the full moon also creates a tidal pull on the molten core of the planet, affecting the incidence of earthquakes.

Why shouldn't other, larger, planetary bodies exert a distant influence on the subtle tides that affect human behaviour here on Earth.

If you still think there is nothing to this beyond a couple of coincidences, hang on, there's more to come.

A lot more.

THE RAINBOW WARRIORS

Visionaries amongst the Medicine Societies of the Native American tribes have their own perspective on this. They too believe we are walking in the last days and although some give us a glimmer of hope, it is only a faint one, dependant upon us changing our ways and learning to live in harmony on and with Mother Earth. One Hopi prophecy involves The Great Day of Purification which will culminate in either:

> "**total annihilation, or total re-birth. The choice is ours. War and natural catastrophe may be involved. The degree of violence will be determined by the degree of inequity caused amongst the peoples of the world and in the balance of Nature.**"

Another prophecy came as a powerful vision to Black Elk, a Lakota Holy Man, and it parallels the Book of Revelation in many ways. The actual details are quite lengthy but in WARRIORS OF THE RAINBOW by William Wiloya and Vinson Brown, its meaning is summarised as follows;

> "**The central meaning of Black Elk's dream is very clear, for it is repeated over and over in different ways. The people of the world shall go through a very bad time, a time of wars and troubles, a black road time. Then they shall be awakened; their hearts will be enlightened and they will work to bring a good time to the whole earth, a time when all men learn to gather power and goodness from god. Here indeed is a wonderful vision of the future in which all the people will be gathered in one fold and one shepherd's flock, when all the many religions will become one big religion with nothing narrow about it, big enough for all and there will be no war any more.**"

The title of Wiloya and Brown's book comes from another tradition, that of the Rainbow Warriors, people of all races and all colours, who will come together in the end times to heal the Earth and bring about the dawn of a new age. Jamie Sams, a native American visionary teacher speaks of them as:

> "the product of a thousands of years of melding among the five original races. These Children of Earth have been called together to open their hearts and to move beyond the barriers of disconnection. The medicine they carry is the Whirling Rainbow of Peace, which will mark the union of the five races as one."

At this point, we begin to stray into New Age territory, simply because Native American traditions which embrace an ecologically sound respect for Mother Earth and encourage "walking in balance with nature" have inspired much of what is now considered to be New Age thinking, particularly in the USA.

Eagle Man Ed McGaa's book THE RAINBOW TRIBE is a prime example. Ed McGaa is an Oglala lawyer, writer and teacher. It is obvious from his work that he believes the world is approaching a crisis point and that the attitudes inherent in Native American spirituality can be part of a relevant solution to the spiritual and ecological problems created by a purely materialistic world view. It is also obvious that he considers many of today's New Agers to be part of the Rainbow Tribe, or the Wigmunke Oyate, as it is known in the Lakota language.

NEW AGE OR ANCIENT WISDOM ?

Jose Arguelles' original books about the Mayan time cycles were also incorporated wholeheartedly into the New Age movement and initiated the world-wide Harmonic Convergence day of prayer and meditation in August 1987. In fact, so many facets of New Age philosophy are borrowed wholesale from a diversity of traditions, it is actually not very new at all, merely a re-cycling of ancient wisdom, most of it reflecting a holistic or Hermetic philosophy, with a little fringe spiritualism thrown in for good measure.

As the Age of Aquarius, it does have some foundation in a natural cycle of the heavens mentioned in the Vedic scriptures and known to the ancient Egyptians: the Precession of the Equinoxes, or the Great Year. This involves the apparent movement of the sun through all twelve constellations of the zodiac, when observed at sunrise on the spring equinox and is the result of several interacting factors, including the axial tilt of the Earth.

The entire cycle takes 25,776 years and the equinoctial sun takes 2,160 years (one of those numbers again) to pass through each zodiacal sign. In this system, we left the Age of Pisces and entered the Age of Aquarius in around 1967, or the summer of love, as the Flower Power people of the time christened it.

My interest in the Age of Aquarius and the new age generally, escalated enormously after the dream I described in Chapter 3, in which St. Mary's church in East Barnet was described as a spiritual power station. As you may recall, in the dream, the church tower became a fountain and while water poured from it, an angelic vision of the white lady explained;

**"These are the Waters of Aquarius,
sent to wash over the land, to cleanse,
purify and heal the Earth."**

At that time, I hadn't discovered a single one of the Earthstars patterns and had no inkling of any connections to biblical prophesies or any others, yet I'd been given a clue that what I was being drawn into was linked to the dawn of the Aquarian age.

As the Earthstars discovery unfolded, I found many of the concepts embodied by New Age philosophy were remarkably appropriate to Earthstars quest. The fundamental premise underpinning the whole movement is the paradigm of a holistic, living, universe. Sir George Trevelyan, one of the most respected visionaries of this century explains simply in his book, A VISION OF THE AQUARIAN AGE;

" **The spiritual world view is a vision of wholeness, an apprehension of the essential unity of all life.**"

Blake put the point a little more poetically ;

" **Every rock is deluged with deity.**"

This viewpoint eludes some people, so perhaps it requires a certain visionary perspective, nevertheless, it mirrors the beliefs of ancient civilisations and modern pantheists, the beliefs of the Druids and early Celtic Church, Native Americans and indigenous tribal societies the world over, as well as recent scientific theories like Dr James Lovelock's Gaia hypothesis. "**All things are connected**" is a one of the tenets of almost every esoteric tradition, as well as Native American beliefs.

Eagle Man Ed McGaa expands on it in his book MOTHER EARTH SPIRITUALITY:

> **"The Sioux were taught and understood that all things are of the Great Spirit. Trees, rivers, mountains, grass, four-legged animals, two-legged animals and winged creatures all came from the Great Spirit, Wakan Tanka, who is the supreme being."**

Our own Celtic heritage is imbued with the same vision. Phillip Carr-Gomm, The Chosen Chief of the Order of Bards, Ovates and Druids, tells us that;

> **" Druids recognise that all life is interconnected and that all life is sacred."**

With apparitions of the Earth Spirit making unannounced intrusions into my life in the guise of goddesses, white lady figures or Tarot archetypes, I didn't have the option of regarding the idea of a living Earth as a theory I could consider completely objectively. Since being drawn into the Earthstars discovery, it was a view of the world I was experiencing personally. I had no choice but to accept its reality wholeheartedly.

THE SECOND COMING

The New Age perspective on the possibility of an apocalypse, still suggests the end of the world as we know it, but reflects a far more optimistic and positive outcome and one which is startlingly relevant to the Earthstars discovery

The works of two of the foremost prophets of the New Age, Ken Carey and the late Sir George Trevelyan, are underpinned by the concept of the Earth and Universe as a spiritual entity of which we are all an individual part. In SUMMONS TO A HIGH CRUSADE (1986), Sir George Trevelyan states;

> **"Remember that the holistic picture implies that everything is organism and alive; the Earth is alive; and you don't get a living organism in the middle of a dead mechanism."**

"This new age movement is not someone's thought out plan for a better society. It is a spiritual awakening to the reality of something greater beyond us."

Sir George Trevelyan's overview of any imminent ecological disaster is that; **"the problem is already insoluble by human sufficiency"** and we need help from the holistic universe.

In his opinion, we've made our planet a sick part of the organism and this has provoked a healing response from the universal immune system which may turn out to be a necessary part of our evolutionary progress.

Another way both he and Ken Carey interpret this is as the second coming of Christ, but not in the obvious way. Both firmly believe that the second coming does not hinge around Christ returning as a man, but as a cosmic energy, a divine impulse which enters the hearts, minds and souls of the human race through a union with the collective soul of the Earth, the Earth Spirit.

Could this be the union hinted at in the book of Revelation and graphically depicted in the Earthstars geometry of the New Jerusalem?

Didn't the symbolism indicate a marriage between the Lamb of God and the Lamb's Bride, the Earth Spirit ? It would certainly be a virgin birth; through the impregnation of the World-Soul by a cosmic impulse, a new energy which would uplift and enlighten all those attuned to it.

If this is the second coming, Christ would not be born of woman, but of the Earth Spirit, an energy body that encompasses the entire planetary populace and penetrates every single heart mind and soul.

Sir George Trevelyan, expands on this theme in more detail in SUMMONS TO A HIGH CRUSADE ;

> "An operation to heal the planet has been launched. For the last ten years, a power which is best called love - a very high frequency of light and harmony - has been pouring into the world of matter and raising its frequency rate. As human beings we are part of the world of matter, our frequency rate is rising. We are on the threshold of fourth and fifth dimensional consciousness. Change is taking place."

In the foreword to MERLIN THE IMMORTAL, produced by Peter Quiller, Courtney Davis and Michael Joseph, we find Sir George expressing exactly the same sentiments, but with a few interesting additions;

> "In our age, an energy that is light, life and love will flood the Earth, re-animating the realm of matter, raising its vibratory rate to that of the spiritual worlds. The two will interpenetrate and interact. So high is this frequency that it will repel all particles, energies and beings attuned to the lower frequencies of egoism, greed, selfishness, violence, hatred, rivalry and war. There will be a total transformation."

These extraordinary assertions speak of an increase in the vibratory rate of sub-atomic particles, suggesting that in the music of the spheres, the Earth will sing in a higher key.

Extraordinary or not, it is of enormous relevance to Earthstars because the way this is supposed to happen is through the Earth's power centres, our sacred sites.

These pulses from the greater soul of the universe are expected to filter down to Earth firstly through major power centres where the Earth's energies connect into the greater universal network. They permeate the Spirit of the Earth through the web of life which links the sacred sites of all religions and, since we humans are part of the Earth and part of the collective World Soul, those energies also influence us.

A network like London's Earthstars, a complex pattern of important sites in a specific design reflecting a clear and definite link to the cosmos, would be exactly the sort of device that would be expected to carry these transformatory impulses.

Several other authors speak of ancient sacred sites awakening or re-activating as a result of a new flow of energy through them.

In MERLIN THE IMMORTAL, by Peter Quiller, Courtney Davis and Michael Joseph, I found the following reference;

> **"If you were to study a map of England and plot on it some of the power centres that are already known, you would find a design emerging, and from that design you can begin to build a picture that reflects the evolution of the Earth itself.**
>
> **The power waiting to be tapped at these sources - by those of you with unselfish motives, and this I cannot overstress - could generate peace and confidence among the people of the world, thereby adding impetus to the forces of light in this ever intensifying battle between harmony and chaos."**

As you can imagine, it struck several chords with me, not just for the suggestion of a coherent pattern of power centres, but for the added notion that the energy flowing through them might have a beneficial, healing effect upon mankind. Two paragraphs later, I received another jolt. The revelation that this remarkable statement had been made in 1967, fifteen years before the Earthstars discovery began taking shape on my O.S. maps. Even more mind-boggling was the news that it had come from someone called Helio Arcanophus, as a message channeled through a medium who was not named in the book.

It is probably no coincidence that one of the Hopi Indian prophesies contains similar ideas, all relevant to Earthstars, including one which names a date when these sites would activate. THE HOPI PROPHESY, A VISION OF HOPE, published by Acorn Press in 1987, states, on page 22;

> **"1986. When Tagashala and the enlightened teachers begin to open the veil of the crack between the worlds. We will see our memory circles. All kivas and sacred power spots will come alive in 1986 and be totally awakened."**

I didn't give much thought to this until I was trying to figure out why the Earthstars main North-South axis isn't exactly aligned directly North. It is tilted a couple of degrees to the north west. Then, I stumbled upon the realisation that in 1986, it had come into alignment with magnetic north which isn't constant and moves slowly through a recurring cycle. In 1986, when the Hopi prophesy predicted that sacred power spots would come alive, the Earthstars North-South axis had aligned with the Earth's North-South axis. I had no way of finding out, in retrospect, if any new energy had surged along it at that time, but if it had, I would not have been a bit surprised.

Other parts of the same prophesy mention the **"twelve great driver wheels of the eight powers."** The eight powers are likely to be the powers of the eight directions which are so important to Feng Shui and all other geomantic spiritual systems. I suspect that the Earthstars mandala is one of those great driver wheels; a vortex in the Earth's energy system; a planetary chakra, where the Earth's energies link to those of the cosmos. After all, that's exactly what its design represents. If some critical evolutionary impulse is beginning to function through the Earth's life-force, a specific arrangement of sacred sites like London's Earthstars would be exactly the sort of circuitry it would use.

Those readers who have no personal experience of these things are entitled to remain sceptical. I had no such option. Call me naive if you like, but I believe whole-heartedly that this is the explanation behind my dream about St. Mary's church.

In the grand scheme of things, ancient sacred sites like St. Mary's really are spiritual power stations and since the power coming through them is what the bible refers to as **"the waters of life"** which seems to be the both a re-vitalising aspect of the Universal Life Force and its Christian counterpart the Holy Spirit, it can definitely be a beneficial, healing influence.

As power sites, these places may not be as active today as they were. If they are undergoing a process of re-activation, they are far more relevant to the future of the human race than its remote past. Their locations identify many of the major and minor points in the living Earth's spiritual energy system. Some of the larger sites, such as Stonehenge and Avebury are in themselves places of enormous importance, both to their local ritual landscape and to the planet at large. Something on the scale of London's sacred geometry, a complex system of sites arranged in precise geometric patterns covering over 400 square miles, has to be of global significance.

It's a vortex which links the soul energies of the planet to the soul energies of the universe; a portal through which the transformational impulses of evolution flow. The web of life radiating out from it through the starfields, links to other power points and forms part of a global grid around the entire planet. In this, we live, breathe and have our being because we, in turn, are part of it.

How the energies affect us varies enormously from person to person, depending on character, emotional states, psychological and spiritual make-up. It also varies from place to place. But we are dealing here with elements that function largely on a sub-conscious level, connecting all things.

Whether it is the dawn of a new day or a new age, the bulk of the population slumber on blissfully unaware of the new light and energy streaming over the horizon. This situation may well change dramatically in the near future. At the turning point of an age, the energies themselves intensify and undergo a transformation, A new energy, a higher frequency begins to flow as evolution instigates the next step forward for humanity and collective planetary organisms alike.

Sir George Trevelyan tells us this will effectively raise the vibratory rate of every particle, every atom, every molecule, of everything on the planet. What that might entail in terms of human experience I hesitate to guess. From Sir George's suggestions that low-frequency, self-centred motivations will cease to exist, it sounds like it may precipitate a change in the emotional climate.

That in itself is likely to bring a change of perception, so that we become aware of a new dimension to life, or at the very least, begin to see the world in a different light. Whether that extends as far as humanity once again walking the Earth with angels, as some claim, remains to be seen.

If Sir George Trevelyan, Ken Carey, the Hopis, Black Elk, The Mayans and all the rest are right, the transformation has already begun. The agents of change are loose in the land and active in the collective psyche of mankind.

To quote The Hitchhiker's Guide to the Galaxy; **"This planet is scheduled for redevelopment."** All the cosmic signs were clearly displayed. It's our own fault if we haven't noticed them.

I know from the rational, materialist point of view, it all sounds more than a little suspect, if not totally unbelievable.

But quite frankly, I don't think the rational, materialistic point of view should be the deciding factor in these matters. That's what got us into all this trouble.

From a visionary perspective it makes perfect sense.

What we need now is a return to the visionary perspective.

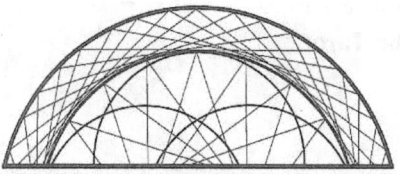

Chapter Sixteen
The Waters of Life

As Gordon Strachan states in his book; CHRIST AND THE COSMOS, the New Jerusalem is a healing influence.

" It is a message for healing, not disease, for harmony, not discord, for revelation, not apocalypse. "

My initial visionary dream clearly indentified the Earthstars sites as a similar beneficial force;

"The waters of Aquarius, come to cleanse, purify and heal the Earth."

This is not a literal reference to water. I think it more likely to be an analogy for the Universal Life Force inherent within the Earthstars geometry. The bible also contains frequent references to the waters of life and these, too, could be symbolic of the spiritual waters, the energies of the Life Force.

"And he shewed me a pure river of the water of life, clear as crystal, proceeding out of the throne of God and the Lamb."　　　　　　　　　　　Revelation 22;Vs 1.

"I am Alpha and Omega, the beginning and the end. I will give unto him that is athirst of the fountain of the waters of life freely."

　　　　　　　　　　　　　　　　Revelation 21 vs 6.

In the Tarot, the water bearer is represented by the card of The Star; so in this ancient esoteric tradition as well as the Earthstars discoveries, the waters of life are carried by the star (or stars). That statement alone is astonishing enough, but there is more. Traditionally, the card is said to represent the Queen of Heaven or Sophia, Goddess of Wisdom

Illustration 169: The Star, card number seventeen of the Tarot Major Arcana.

pouring the waters of life upon the Earth. Again, that is enormously relevant to how the Earthstars discovery was revealed.

The more mundane interpretation of the card usually relates to personal hopes, optimism for the future and the fulfilment of dreams, wishes and aspirations. It's an extremely beneficial card. In TAROT TRUMPS, THE COSMOS IN MINIATURE, John Shephard states;

> **" In Christian allegory, the star in the east was often identified with Venus as the morning star which rose before the sun, Christ; and as the Virgin, The Star carries the promise of the Redemption which will deliver mankind from the Fall shown in the previous card, The Tower."**

Obviously, there are quite a few correspondences between Earthstars and the Tarot card of The Star. Briefly though, if the Earthstars energy or influence can be understood to be related to the **'waters of life'** poured upon the world, they too are an extremely beneficial influence. Revelation stresses that the New Jerusalem is; **"for the healing of the nations"**.

> **"In the midst of the street of it, and on either side of the river, was there the tree of life which bare twelve manner of fruits and yielded her fruit every month; and the leaves of the tree were for the healing of the nations."**
>
> Revelation 22; vs 2.

Certainly, this is part of Ken Carey and Sir George Trevelyan's insights into what kind of transformation of the New Age might bring. To their way of thinking, an input of a different energy into the Earth's sacred sites, in specific places, is likely to be felt as an intuitive, emotional response by many people. It won't be something you think about and consider. You'll just feel it.

What it will change is the emotional climate of entire areas.

This is what will provoke a dramatic change in the way we perceive the world. In all honesty, no-one can really know if any of this is anything other than wild speculation or wishful thinking, but I can't sit on the fence after all I've been through with the Earthstars phenomena. I hope and believe the Earthstars patterns are here to act as circuitry for an force that will transform the Earth by its beneficial healing influence.

I don't expect change overnight. These are, after all, an aspect of the forces of evolution and so may move at a less than glacial pace. One possible perspective on these ideas is that the Earth is re-attuning itself to the greater harmony of the universal life-force to redress imbalances humanity has created here on Earth in the past few hundred years. The mechanics of how this may have a practical healing effect on people and places provides us with an insights into how the kind of hands-on spiritual healing practised in Evangelical churches and spiritualist churches actually works, as well as revealing what Earth energy patterns like London's Earthstars may have to do with it.

This particular revelation isn't something I discovered personally. It was gleaned from a gentleman called Colin Bloy at a lecture he gave for R.I.L.K.O. The Research into Lost Knowledge Organisation. Founded in 1969, the society holds meetings and lectures throughout the year and I have been a member since the mid-eighties. Around 1985, one of the guest speakers was Colin Bloy who was to deliver a talk entitled "Ley lines and the collective sub-conscious." I'd gone along to see if his views and experiences threw any light upon my own. Boy was I in for an evening.

For a start, I discovered that Colin Bloy had a unique combination of talents. He was both a dowser and a healer and this interesting blend of abilities had revealed a hitherto unsuspected link between the two practices.

As well as dowsing energy lines in the landscape, Colin dowsed patients to assess the condition of their personal energies (or auras) and he didn't use bent coat-hangers, twigs or pendulums. He simply felt the energies in his hands, as I and many other healers sometimes do.

Unlike myself, Colin Bloy had developed a great deal of sensitivity and could glean enormous information from what he detected. I was aware that many healers scanned their patients in a similar way. I had been taught some of these techniques myself. What Colin added was an understanding of how the energy transferred during healing, worked in conjunction with the Universal Life Force and the Earth's energies.

Almost without exception, healers maintain that the energy they use to heal comes <u>through</u> them, not <u>from</u> them. They consider it a divine spiritual energy whose source is the creator, God, or whatever you consider to be the common source of all creation.

Since Colin could sense energies in the landscape, he had begun dowsing healers as they worked on patients to see if he could detect this energy being directed through them. He was curious to see what changes the healing process created in and around the healers' body.

It was an astonishing discovery. First, he found that the healer, quite unconsciously, plugged into the Earth's circuitry through a specific pattern. As the healer began working, he or she would become the centre of a cross of energy lines, each arm of the cross extending roughly to the four compass points and linking into the nearest local sacred sites (or power sites) in each direction.

Secondly, a vertical column of energy would then switch on like a tight focus spotlight beam shining down on the healer. The energy from this column of light was what the healer channeled to the patient and as the healing progressed, the column of light enlarged so that it engulfed the patient as well as the healer.

In a flash, I understood that the geometry of the Earthstars mandala was a large scale version of the column of light. Not only was this clear confirmation that the hidden energies of the landscape could definitely be used for healing, it demonstrated how.

I had noticed myself that the sensations I felt at some sacred sites was similar to those I felt in my hands and arms when practising hands-on healing. David Furlong, who is also a healer, makes the same observation in his book THE KEY TO THE ANCIENT MYSTERY.

Colin Bloy's revelations opened up the possibility that these vertical inputs of healing energy exist naturally at places in the landscape which have become recognised as our sacred sites. Indeed that energy and its effects on human emotion and spirit may be what made them sacred in the first place. It also planted the idea that the Earthstars geometry might be a large scale version of the geometry at work within the column of light.

The most important insight though, was that it could be induced consciously anywhere and channeled, by slipping into the same altered state of consciousness used by healers, a contemplative state most easily achieved through meditation.

Colin Bloy had already tried dowsing people during meditation and found that they created the same energy patterns around them. Once tuned in, they became the centre of a four-armed cross in the Earth energies and a single axis of light/power from above. Moreover, he found that, in meditation, the beneficial effects of the column of light were channeled through the meditator into the Earth where they created changes in the dowsable energy patterns around the healer.

THE FOUNTAIN EFFECT

Having made this discovery, Colin realised it might have practical applications. If you could heal individuals with this energy and it had a beneficial affect on the Earth's energy fields, perhaps it could be used to heal places that didn't feel quite right, too: places where the atmosphere felt uneasy, or their were high levels of crime or violence.

This unusual idea was put to the test in Brighton, back in the days when every Bank Holiday brought an invasion of warring gang members. A group of healers and dowsers assembled by the Old Steine fountain (reputed to take its name from a megalithic circle of old stones still beneath it). Their aim was to meditate with the intention of spreading an atmosphere of peace and goodwill throughout the town to defuse the expected bank holiday violence.

Remarkably, the dowsers reported that the meditation had brought about a distinct effect in the town's energy lines, amplifying them so that their dowsable energy was dramatically increased. Even more remarkably, over the next few days, the bank holiday trouble never flared and subsequently the local crime rate fell. Those involved, quite reasonably, considered the experiment a success. Such a success, in fact, that the idea spread to other spiritual healers who then began to incorporate this community healing into their other activities, at first, locally in the Sussex area, then gradually on a more widespread basis. A network of healing groups sprang up and as the first one had been a Fountain Group, that's what the others were called. Since the energy they were working with is synonymous with the concept of the waters of life, the symbolism of the fountain is an extremely appropriate one.

During the eighties, Colin Bloy's lectures spread the idea and it caught on. There are now Fountain Groups all over the world (see www.fountain-international.org). It's a loosely knit organisation, with very little in the way of a structure. It's held together by a core group of unpaid volunteers and local groups who maintain a sense of unity with a regular newsletter and an annual conference.

One or two of the traditional ley-hunting fraternity remain sceptical about such matters and about dowsing in general. Needless to say they are not dowsers, psychics, healers, or particularly spiritually inclined. It seems that to appreciate these facets of the phenomena, you have to have a little of them in your own make-up.

The Earthstars discovery is an enigma with its foundation in our sacred sites. It is a spiritual mystery. It has to be understood from that perspective and dowsing provides a useful link between the concepts of energy, power and spirit. Dowsers have consistently stated that energy and spirit are two sides of the same coin.

John Michell's work, thirty years ago, pointed to lines of spirit, in terms of the Earth Spirit, spirits of place, spirits of the land, all of which may be perceived by the non-spiritual as some kind of barely tangible energy presence.

Since humanity is part of the Earth, we all have connections to the Earth Spirit and can interact with it. That is why the Fountain Group experiment is so relevant.

It is a way of consciously working with the Earth's energy fields to try to give certain areas a positive boost by making a deliberate conscious connection to the greater harmony of the Universal Life Force.

If that sounds like wishful thinking, it is. I'd like it think it works. According to Fountain's leaflet and to brilliant dowsers like the late Hamish Miller (co-author of THE SUN AND THE SERPENT) and Colin Bloy who both worked with Fountain, it does. I accept that it is based on the subjective art of dowsing and is therefore difficult to prove. But in my opinion, if you can make this world a better place simply by meditating for fifteen minutes a day, more people should try it.

The Maharishi always maintained that if just two percent of the population practised transcendental meditation, it would lower the local crime rate. In fact, any form of meditation will achieve the same results. Impressive testimonies to the personal benefits of meditation can be found in every book on the subject.

Less well known perhaps are its effect on the collective psyche. Everything that happens in your mind simultaneously happens in the collective mind of which you are an individual cell. Every single person who re-attunes to the divine within themselves, not only begins to re-attune, replenish and re-balance their own energies, but those of the Earth and their local environment. In short, meditation's capacity to plug you into the collective subconscious and the Universal Life Force makes it a powerful tool for planetary transformation as well as for personal enlightenment.

"Work on yourself and serve the world" was said to be the motto of the grail knights. As it turns out, the two tasks are inseparable. When you do one, you automatically do both. As you change yourself, so you change those around you.

This leads to another of the most important grail secrets; that the land and the people are one. As you change yourself, so you change the world. Revolution is always an inside job.

Until I stumbled upon this realization, I'd often wondered how the meek could possibly inherit the earth. I gave them a snowball in hell's chance against the guys with AK47s and Uzis.

In the light of all this, maybe they will just sit and quietly meditate for a few minutes each day, joining forces with the incoming energy as it increasingly takes effect on those around them, slowly making the world a more peaceful place.

Meditation applied specifically for these ends does for peace what sneezing does for the common cold. It makes it highly contagious affecting everyone and everything in our immediate vicinity. Once you appreciate these natural interactions with your psycho-spiritual dimensions, you become an active agent of transformation.

Of course, there is also an enormous amount of more practical work to be done. Sitting on your backside meditating isn't going to replant the rain forests, but as you well know, in the right atmosphere everyone is motivated to work better, think more positively and co-operate more effectively. The popular term "healing the Earth" isn't altogether appropriate.

As a self-regulating organism, the Earth is quite capable of healing herself and left to her own devices, she would.

Unfortunately, she has millions of us humans teeming all over her making a mess of things. What needs to be healed, is our relationship to the Earth, our Mother planet.

Earthstars is not simply about the connections between sacred sites. It is about the hidden connections each one of us have to each other, to the Earth, to the universe at large and to the creator of it all.

To fully understand these connections, we have to re-make some of them.

We have to plug ourselves back into the cosmic mains.

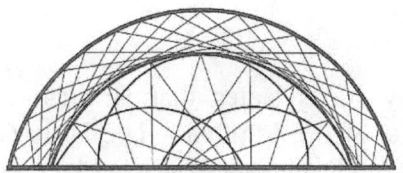

Chapter Seventeen
Camelot, The Round Table and The Holy Grail

Ever since the first apparitions of the Gwenevere look-alike at Camelot moat, the Arthurian legends have held a fascination for me, but not simply as entertaining tales. In the Western Mystery tradition, they are not to be taken at face value. They represent an ancient belief system incorporating an understanding of the stories' main characters as personifications of natural forces within the land.

Some of the tales re-enact the cycles of nature quite obviously. Sir Gawain and the Green Knight is a clear representation of the annual alternation of supremacy between the winter king and the summer king. Both are personified aspects of a male, solar energy; one the summer sun - god hero who brings growth and life; the other, the winter king who brings death because the season over which he rules sees a natural dying off of vegetation. He alone lives through it as the evergreen knight. Different versions of this myth are found world-wide, but a particularly apt one is the Egyptian legend of the struggle between Osiris, who represents both the Sun and the corn god, and his brother Set, who cuts him down and scatters his remains throughout the land.

As Lewis Spence points out in THE MYSTERIES OF BRITAIN, the grail had its forerunner in Cerridwen's Cauldron of Inspiration and;

> **"The whole legend of the Grail, although diverted to Christian uses, is indeed entirely derived from sources which may well be described as Druidical."**

In Christian terms, the Grail Mysteries are actually the mysteries of the Holy Spirit personified as The Lady to whom the knights must pledge their allegiance and swear to serve. This clearly represents the continuance of a Goddess tradition originally associated with the druidic notion of the Holy Spirit, the nwefre or life force. What really fascinated me though was Bride's connections to the grail legends. THE ANCIENT BRITISH

GODDESS by Kathy Jones states that Brigit's (Bride's) most famous talisman is **"the Grael of the Maiden - The Holy Grail of Innocence of the Virgin Goddess".** Gwenevere was clearly created in the image of Bride, the White Goddess. In Welsh, the name Gwenevere means white spirit, white ghost or white breath of life. Bride, Gwenevere, the White Goddess and the Life Force are interchangeable. It seems the White Goddess, Bride, was at the basis of the Grail legends, as well as the Earthstars discovery and was a personification of the Universal Life Force.

Earthstars second connection to the grail sagas is, of course, the Camelot site, which spurred on my curiosity and led first to the discovery of the Barnet Triangle, then to the whole Earthstars network. It had always seemed likely to me that Camlet Moat, whatever else it had been in the past, must have been a place sacred to the White Goddess and the visions associated with the site were manifestations of her as the genius loci.

Bearing in mind that the grail legends incorporate associations with a pre-Christian goddess religion, I began to suspect that whoever named this place Camelot must not have done so arbitrarily.

If indeed this was an important place where the White Lady was revered and the grail legends do represent a discreet continuation of her tradition as the lady whom the knights serve, perhaps those who were privy to this secret doctrine had named the place Camelot because it played an important part in these mysteries. Perhaps to them it really was a grail castle location.

What I can say for sure is that the mediaeval counterparts of the Grail Knights were said to the Knights' Templar and Camlet Moat has some unusual associations with the Templars.

Sir Geoffrey de Mandeville, the earliest candidate for naming the place Camelot, had a strong, yet strange, relationship with the Templars. When dying, he was visited by a mysterious delegation of Templars. On his death, no one would take his body for burial on holy ground because he'd been excommunicated, yet it was claimed by the Templars, who for reasons best known to themselves, hung the corpse in an apple tree within the temple grounds for 19 years before giving him a decent burial. His final resting place is in the circular Temple church off Fleet Street, amongst the most honored knights of the order; an end which suggests that if he wasn't a Templar, he was held in exceptional high esteem by them. This isn't the only strange story associated with de Mandeville's demise. Another is a folk-tale, obviously with some factual basis since it stems from his disagreements with the King of the time, Stephen.

This story claims that whilst fleeing from the King's men on the night of a full moon, Sir Geoffrey de Mandeville hid at Camelot Moat in a hollow oak. At midnight, he emerged from his hiding place and fell to his death down the well.

A number of inconsistencies are highlighted here, not the least being that Sir Geoffrey actually met his end under entirely different and well-documented circumstances at the siege of Burwell Castle in 1144. So what does this tale really mean?

The mystical elements of druid's oak; midnight (the witching hour); the full moon, traditionally a time for occult rituals; a holy well with pre-Christian goddess associations and a death that wasn't a death, combine to suggest something else altogether.

To me, it sounded like an allegorical tale, not of an actual demise but of a symbolic one; a shamanic death; an initiation.

In particular, it suggests an initiation by drowning, which is in all probability the basis of the Christian baptism. Several authors have suggested that such initiations took place at the Chalice Well in Glastonbury, where a large niche in the stonework beneath the surface would allow total submersion.

The purpose of such an initiation would be to trigger a profound spiritual experience. Many drowning victims claim that the event prompts their whole life to flash before them, or they undergo a near-death experience, witnessing the reality of the after-life personally.

If an initiation or baptism was meant to provoke that effect, it's hardly surprising that it would result in a life-changing experience. The participants would not only feel they have been to the brink of death and been reborn, they may have actually had the rare chance to re-assess their lives from that perspective, or even have glimpsed, for a brief moment, the reality of the spiritual realms.

From this point of view, this folk-tale is not the nonsense it at first seems. Perhaps it is a distant memory of Camlet Moat as a place of initiation into the mysteries of the White Goddess, which are the mysteries of life and death.

Whether the place was called Camelot at the time, we'll never know. In this context, it's pretty irrelevant. Since these activities echo very ancient rites, they would certainly have pre-dated our known associations with the name Camelot unless the Camelot of Arthurian legend was actually named after this location rather than vice versa.

It seems unlikely, I know. But equally, there is no overwhelming reason to rule out the possibility completely, either. According to local historians, the most likely source of the Camelot Moat's name was Sir Humphrey de Bohun who is said to have fortified a Manor House on the site and who occupied it at a time when the Arthurian legends were at the height of their popularity. Yet even he was a descendant of our Norman conquerors and had family connections to the Templars. So his motives for naming the spot Camelot are open to as much speculation as anyone else's.

Some historians link the derivation of the name Camelot to a Celtic god of war called Camulus. That may be so. The idea that it could also be a sacred site associated with a goddess does not necessarily conflict with this.

What better seat of power for the defence of the realm than a sacred isle under the protection of the White Goddess, the genius Loci of Britain, herself ? Whether Camelot Moat gained its name from Sir Geoffrey de Mandeville, his descendant, Sir Humphrey de Bohun, or at an earlier date, we'll never know. Certainly the name Camelot has been definitely be linked to the place for at least six hundred years and according to local legend, possibly for nearly a thousand years or more.

The fact that this particular Camelot has not become more widely known only prompts the question: why? Its anonymity could not have resulted from the place being overlooked, unknown or of no significance. Geoffrey de Mandeville was a larger than life character, a member of the ruling classes and known personally to royalty He was one of the most important men of his time and the location given to him as his manor would have been commensurate with his position.

When the entire area around Camelot was decreed a royal hunting ground, Enfield Chase, Camelot was at its centre, a further indication that it would have been familiar to the ruling classes and the upper echelons of society during that period. It was always known to Royalty, used by them or owned by them. In fact, the daughter of Sir Humphrey de Bohun, Mary, married Henry Bollingbrook who in 1399 became Henry IV.

Sir Walter Scott was also aware of the place some centuries later. He used it as a location in one of his Waverley Novels. Scott, it has to be said, incorporated Arthurian themes, the Templars and references to the Rosicrucian mysteries into his works, so the esoteric tradition incorporated into the Arthurian Legends would have been familiar to him and the significance of the name would not have eluded him.

If Sir Walter Scott knew of it, many of his peers and contemporaries interested in esoteric matters probably did too.

What puzzles me most is this: I would have thought that any place actually bearing the name Camelot would have attracted an enormous amount of comment, speculation and public attention, particularly in the periods when interest in the Arthurian Legends underwent a resurgence or interest, as in the pre-Raphaelite period. So why didn't it? Why has this Camelot been overlooked and ignored for the last 600 years ? Could it be that those who knew about the place didn't actually want to draw attention to it? Could it be that they actually wanted this particular Camelot to remain unknown and therefore, the less people who knew about the place the better? Little surprise then that a clue to its location should eventually turn up in the works of Charles Williams, a known member of an esoteric society, The Hermetic Order of the Golden Dawn, which worked with the Western Mysteries.

As we saw in the early chapters of this book, the map drawn by Charles Williams, placed Camelot in London and depicted the spirit of the land as a Goddess he called Brisen, actually another pseudonym for Bride.

In my opinion, this presents too many coincidences to be a coincidence. At the very least, it suggests that Williams may have been aware of this Camelot site as well as its connections to a cult of the White Goddess as spirit of the land.

It is unlikely that one member of an esoteric society would be in possession of this kind of knowledge independently. Far more likely is the conclusion that the place was an important location related to some of the Society's better kept secrets. Even if, as Gareth Knight suggests, Williams simply regarded London as Camelot, it begs the questions why. The possible involvement of an esoteric society also prompts a new perspective on Enfield's Chase, the Royal Hunting Ground.

In occult symbolism, the Royal Chase relates to the passage of the sun across the heavens, pursued by the moon goddess as Diana or Artemis, the huntress. So is it significant that at the centre of the Royal Hunting Ground was there a shrine to the moon goddess, Diana the Huntress?

If the main lodge of the Royal Chase was Camelot Moat, was it a hunting lodge, a masonic lodge or an occult lodge ?

In a broader context, it is worth pointing out that our aristocratic ancestors had a tendency to landscape their grounds along geomantic lines in an attempt to create their own Arcadias, often with classical temples which we describe these days as follies. It demonstrates a clear awareness of the science implicit in the Earthstars' discovery and suggests the follies were actually real temples, frequently used.

Audley End house is a good example. It has a folly in the form of a

Illustration 170: Rags tied in a tree over-hanging the remains of Camelot Moat's well indicate it is still regarded as a sacred site by some. Similar rag trees may be seen beside holy wells in Cyprus, Eire, Wiltshire, Cornwall and elsewhere.

Illustration 171: The temple folly at Audley End House near Saffron Walden.

Greek temple perched upon a hill overlooking the house. Its siting qualifies it as an active sacred site, rather than a folly.

In THE GREEN STONE by Graham Phillips and Martin Keatman, the follies at Biddulph Grange were said to have actually been used for rituals conducted by former owners of the property and their friends.

If follies like these were really built as temples, what did members of our aristocracy get up to in them? If they laid out their formal gardens in geometric designs based on the very same principles as the Earthstars[7] patterns and the sacred geometry of the great Cathedrals, they must have been aware of its significance.

Were they all secretly followers of a spiritual tradition based on an understanding of Hermeticism and the classical mysteries which were the basis of the Masonic and Rosicrucian esoteric societies to which many appear to have been members?

CAMELOT AND THE NEW JERUSALEM

The fact that Camelot Moat has connections to St. John's New Jerusalem makes it doubly interesting. Like the New Jerusalem, Camelot is a dream of the perfect place, or perfected state of being. It was the mediaeval vision of the perfect city or society, as the New Jerusalem was in St. John's time.

Camelot was the New Jerusalem of its own era.

Both, of course, had a fore-runner in the Temple of Solomon which was the foundation of old Jerusalem.

What I now began to wonder was this: was the knowledge of the Earth Spirit's energies, sacred geometry and use of the Universal Life Force, known to a powerful group in the past who could have identified this pattern (and others) in the Earth and marked just enough of its sites for it to lie undiscovered until now ?

Certainly, as THE SECOND MESSIAH by Robert Knight and Christopher Lomas shows, the sudden spate of cathedral construction throughout Europe in the middle ages was orchestrated by families linked to the Templars.

These buildings employed a vast knowledge of sacred geometry in their design and were positioned at key locations in the landscape, probably to plug into the natural power points of the land. Knight and Lomas also suggest fairly conclusively that the publication of the Arthurian Sagas, which appeared around the same time, may also have been instigated by the same sources, as part of the same grand plan.

It made me wonder if there was a hidden agenda behind the Norman invasion of Britain. This was the Isle of the White Goddess, and she is the Goddess of the Grail legends. Is that what attracted the Norman knights and nobles ? Is there some special power in the land here, represented by the White Goddess?

Is the Grail or the Grail Castle actually here in Britain ?

One interesting insight into this theme was pointed out to me by Peter Quiller, author of MERLIN THE IMMORTAL and MERLIN AWAKES. Circular patterns of power sites, he suggested, might represent a round table emblazoned upon the landscape.

In the Arthurian sagas, the round table was part of Gwenevere's family heritage, part of her dowry, so it is linked more to her than Arthur.

Moreover, it is a symbol of harmony amongst the knights some of whom, as we have seen, in the secret tradition represent aspects of the forces of nature within the land.

The Earthstars patterns are also a symbol of harmony and represent the forces of nature within the land held in a pattern of dynamic equilibrium. In many ways, they are also a table of the heavens, showing the order of the cosmos.

Is the Earthstars mandala, a round table ?

Or is it something even more important ?

The Grail chalice is always held in the possession of a maiden, more often than not, one who represents the White Goddess.

Is it the legendary vessel that carried the life blood of Christ or the chalice that contains the waters of life, the life-force of the universe.

To ask the Grail question: Whom does this Grail serve ?

In the context of the Earthstars discovery, it is the vessel of the life-force. It serves all.

The Earthstars mandala is a structure of the life-force that serves all and resembles a chalice-like vortex in the planetary energy systems.

It is also a vessel through which the waters of life flow.

Is it the basis of the Holy Grail ?

If this assumption is correct, the quest for the Holy Grail in the service of the lady, was actually a mission to free the feminine spirit of the land, the secret goddess, who embodied the invisible Universal Life Force.

In these terms, the Grail would not only give the knight who understood its mysteries the secret of life but also had the power to rejuvenate the Fisher King's wasteland.

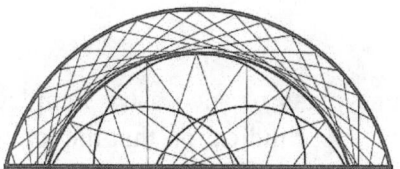

Chapter Eighteen
The Earthstars Vision

> "All ancient civilisations were concerned with proportion - by whatever name they may have called it - for the mark of a civilisation is its concern for universal principles of order and the relationship of people to their universe."
>
> Keith Critchlow

The question of what the Earthstars patterns are has many answers. Inherent in the idea of finding a previously undetected and unsuspected pattern of any sort, is the notion that the pattern implies hidden meaning and that, in finding it, you also find a new and deeper understanding of whatever the pattern applies to.

The Earthstars patterns are a phenomenon which has its foundations in the Earth and in our sacred sites, places that are supposed to connect us to our spiritual dimensions and which bind us to the mysteries of life in general.

Hardly surprising then that various interpretations of the Earthstars geometry point to a hidden aspect of existence which may be both energy and spirit and whose hidden patterns, if we are to believe the available evidence, indicate a coherent divine plan underlying creation, a holistic, living universe and the role of our sacred sites within it as very real spiritual power centres.

Symbolically, the sacred geometry alone represents recurring foundational patterns upon which the structure of the universe and the material world is built. Those patterns can be found everywhere from the sub-atomic scale to the galactic. They incorporate the numerical ratios and proportions of the Earth, the Sun, the Moon and other planetary bodies because they underlie the structure of the entire cosmos and are in a holistic relationship with it.

This is incredibly ancient knowledge, demonstrably employed by the builders of Stonehenge and the pyramids who clearly understood these patterns and their significance. The presence of these specific geometric forms within the ground-plans of successive temples from Stonehenge to the great Cathedrals is a clear indication that the knowledge intrinsic to the Earthstars mandala was possessed by influential elements within the orthodox religions of all ages.

It is not lost knowledge. It has been deliberately concealed, possibly as part of a division of spiritual knowledge. A secret doctrine for some of the clergy and ruling elite. Something else, possibly dangerous knowledge, for the uneducated masses. Unfortunately, it has been concealed so successfully that it is a complete mystery to most of today's priests and to the population at large. The mediaeval church of Rome managed to behave quite hypocritically in this regard. It guarded these secrets so jealously that anyone remotely suspected of dabbling in such matters was declared a heretic for whom a foretaste of hell-fire could quickly be arranged prior to their imminent departure for the real thing.

Under the circumstances, it is hardly surprising that this knowledge was held amongst the secrets of esoteric societies particularly those following the Hermetic tradition. Even in pagan times this was not knowledge to be shared freely with the uninitiated who may not appreciate its worth. It was for the initiated only and was without doubt an important element with the teachings of many mystery schools. In THE CANNON, William Stirling makes this comment on the Pythagoreans;

> **"The Pythagoreans concealed their doctrines within a numerical and geometric system which was the only form of their philosophy to be given to the outside world."**
>
> **"Plato wrote over the door of his academy;**
> **Let none ignorant of geometry enter here."**
>
> **" From this it may be concluded that Plato meant to inform us that no-one could understand his philosophy without knowing the geometrical basis of it, since geometry contained the fundamental secrets of the ancient science."**

The evidence would seem to indicate that the Earthstars patterns derive from humanity's earliest mystical sciences and from an energy, or energies, which can be described in a number of ways, all of them equally astonishing: the Universal Life Force: the Forces of Creation.

Or in Christian terminology, the Holy Spirit. If this is correct, the fundamental secret of the ancients was a science of the Holy Spirit. An art likely to have encompassed an awareness of the Universal Life Force, its interactions with the personal energies of the individual human aura, the collective psyche and with the Spirit of the Land. Clearly it involved harnessing, manipulating and amplifying spiritual power through the use of geometric harmony in architecture and probably on a grander scale in the landscape at large.

GEOSOPHY

By its very nature, the Universal Life-Force is omni-present and theoretically accessible to all through a form of gnosis or direct personal relevation, particularly at places which might be regarded as its power centres - the Earth's sacred sites.

In the light of this information, how the Earthstars discovery came about is not as extraordinary as you might at first have assumed. Dreams, visions, revelations are the expected results of encounters with our spiritual dimension at places where we traditionally commune with our god, or gods and goddesses.

Why the apparitions of deific figures took feminine form as opposed to the more convential idea of God as a paternal archetype isn't so extraordinary either.

The forces they represent are all traditionally perceived as a nurturing female influence - Mother Nature - Mother Earth - the Earth Spirit - the Holy Mother Mary. In Hinduism, the active principle of the divine is feminine, Shakti. Even the Holy Spirit, the active priniciple of the God seems to have been an inherently feminine manifestation concealed as one of the great mysteries of the Christian Church. The Universal Life Force can be all things to all people. It is part of us and we are part of it. Visions of gods, goddesses, mythic figures or biblical characters may be the result of an interaction with its component energies on a deep subconscious archetypal level which creates anthropomorphic forms we can individually recognise.

The way in which the Earthstars discovery was initiated, through a series of dreams, visions and psychic insights, could be the result of these subconscious processes and might be described as a form of gnosis, knowledge of the divine through direct revelation. A slightly more apt term might be Geosophy, meaning literally "the wisdom of the Earth." This is a little more relevant to Earthstars in that its etymology stems from the combined names of two classical goddesses, Ge or Gaia, the Earth Goddess and Sophia, Goddess of Wisdom, both of whom are implicated in the revelatory process which led to the discovery of the Earthstars patterns.

I have always maintained that this geometry was not a discovery in the usual sense of the word. In my original Earthstars book, published in 1990, I said that I regard it as a release of knowledge from the Earth herself. My opinions on that have not changed at all.

The forces inherent in the Earthstars network are both energy and spirit and can be comprehended as either. The patterns themselves are as much a manifestation of the Earth Spirit as the White Lady apparitions. Many of the key sites in the Earthstars mandala are places where the active force has been perceived throughout the centuries as an essentially feminine influence, initially, as Mother Earth, or Bride - the White Goddess, then later as Mary, the Divine Mother. In other countries and cultures she was known by many different names and had many differing aspects, so she obviously represents a recurrent archetypal principal.

She may be considered to be the spirit of this land, the soul of the land, the spirit of the Earth, an aspect of the Holy Spirit or as an anthropomorphised manifestation of these forces. With the coming of Christianity to these isles, initially under the auspices of the Celtic church, which embodied much of the druidry it replaced, many of her sites were re-dedicated to the Holy Mother Mary.

It is important to remember though, that this is not an exclusively Christian mystery or an exclusively pagan one. The Earth's sacred sites are not sacred to just one religion or any one group who happens to own them at a particular moment in time. They are sacred to all people. The mystery underlying them is a unifying factor common to all religions and has its roots in the natural structure and order of the cosmos and the source of all creation. All things are connected. What connects them is an invisible web of energy and spirit. We are linked to each other, to the Earth, to the cosmos and to our creator by the invisible strands of the Universal Life Force of which we (and all other things) are a part. Thus behind all religions is a hidden unity kept from us by disparate belief systems which all too frequently add a divisive element, by insisting that their one particular understanding of these mysteries is the only valid one.

For instance, many Christians will accept the notion of the Holy Spirit, but may have their beliefs challenged by any suggestion that it is remotely connected with the subtle energies of sacred sites and ley alignments, or with apparitions of "White Ladies".

Devout Catholics may be more amenable to white lady visions like those witnessed by Bernadette Soubirou of Lourdes, as long as they are identified with the Virgin Mary, but may not welcome the suggestion that such visions are a non-denominational phenomena with other white ladies being interpreted as ancient goddesses like Bride, Diana, Cerridwen, Isis, Sophia, Athena or their North American equivalent, the Lakota's White Buffalo Calf Woman.

All these may be merely different outer expressions of the same force or essence, varying from culture to culture, or from place to place and dependant on local interpretation, just as there is one creator who is painted in different guises by different religions.

The Earthstars mandala is an extremely complex, multi-faceted phenomenon. Not only can it be appreciated as symbolic geometry, energy and spirit, it may also be understood as St. John's vision of the New Jerusalem in the Book of Revelation.

The New Jerusalem is frequently described as the Heavenly City or Celestial City, terms that could be taken to mean it is a symbol of the spiritual dimensions, the perfect essence behind reality rather than a particular place in the physical world. In this sense, it is a construction that has its foundations more in our spiritual realms than the material world. Even more remarkable, in my opinion, are the Earthstars' connections to the legends associated with the Holy Grail.

The Earthstars' quest began at Camelot. It involved Gwenevere-like apparitions. The overall geometry is comparable to a round table and as the Universal Life Force, is the essence of the Holy Grail. (Like the Grail, the life force serves all and is the spiritual life blood or Sang Real or true/real life blood that sustains all life. A quest for either one is the eternal quest for the mysteries of life itself.

Reassuringly, all the various aspects of the Earthstars patterns have one thing in common.

They are all simply different ways of explaining the same thing.

HEALING THE WASTELAND

The purpose of the Earthstars' energy patterns, if we are to believe the meaning of my dream, is to heal the Earth through a release of the waters of life. This has a direct parallel in the Arthurian legends where the grail has healing properties and possesses the power to restore Arthur's Kingdom and heal the wasteland. Arthur's Kingdom is, of course, Britain and its restoration implies a return to a Golden Age of perfection. The restoration of the Kingdom is also a Christian concept relating to the coming of the Kingdom of Heaven on Earth. Surprisingly, it was also the basis of much esoteric activity.

In THE EARTH SPIRIT, ITS WAYS SHRINES AND MYSTERYS, John Michell informs us that seventeenth century verses by Stolcius; **"Refer to the grand and oldest form of alchemy which was concerned to bring about the earthly paradise through a fruitful union of cosmic and terrestrial forces."**

So what is the wasteland? In this context, the wasteland represents a spiritual desert, a purely materialistic view of the world. A world with no spirit. A dead land. Sadly, that is how many people view the world today. The wasteland is all around us, built by property developers, town planners, architects and designers who have no concept of working to a divine plan and are motivated by a desire for maximum profit rather than the idea of creating sustainable communities in harmony with the natural world. You see it all around London, typified by developments where as many flats or homes as possible are crammed into any available space, usually with little or no garden and not even a grass verge on the road outside.

In a broader sense, it is anywhere created, populated or governed by those whose values are atheist or materialist and who have no spiritual dimension whatsoever in their lives. As such, the wasteland is a dreadfully negative example of the grail secret that the land and the people are one. The legacy of the wasteland is death, for it is a land sowed with the seeds of its own destruction. Those who see the world as an inanimate rock are in serious danger of turning it into one. An insight into this can be gleaned from the foreword to A HOPI PROPHECY published by Acorn Publishing in 1985.

> **"We are dreamers and we create physical reality through the thought-forms we hold. As things stand now, the human race is dreaming a hell. But as individuals and as a collective humanity it is our potential to dream of a paradise and to create the kingdom of heaven on our Mother Earth."**

The world as we know it today is a joint enterprise between humanity and the creator. Unfortunately, mankind's contributions do not generally reflect any awareness of a partnership of any sort, least of all divine. They demonstrate a short-sighted, self-centred attitude that is frequently pitted against the natural world rather than working in harmony with it. A sense of unity with nature is conspicuously absent.

There are a great many people who for a variety of reasons, are busy creating hell on Earth; drug dealers; street gangs; vandals; muggers; organised crime syndicates; corrupt politicians; political and religious fanaticists; some of the big corporations which are like vast machines with no heart or soul. The list could go on. The common denominator is their lack of concern for anyone's interests but their own. The restoration of the wasteland implies a spiritual vision of the land as an aspect of the holistic universe, animated by the Holy Spirit or Universal Life Force and sharing its spiritual dimensions with its populace.

Our spiritual dimensions are not just within our own individual minds, they also also within the collective psyche and planetary mind-fields. They are out there, in the world around us.

The world is a different place for those who have had the greater reality of the spiritual realms revealed to them. For them, as William Blake said: "**Every rock is deluged with deity.**" He meant that the world is alive, animated by the divine spirit of its creator. Blake, of course, was a life-long mystic who conversed with angels and other disincarnate spirits on a daily basis and was convinced that he was "**under the direction of messengers from heaven**". He was also The Chosen Chief of The Druid Order from 1799 until his death in 1827. He would have been well aware of the visionary landscape and, like St. John, glimpsed the reality of it as a vision of the New Jerusalem. Blake's vision though, unlike St. John's, had a definite physical location, London.

> "**The fields from Islington to Marylebone to Primrose Hill and St. John's Wood; were builded over with pillars of gold and there Jerusalem's pillars stood.**"

It is remarkable that William Blake should have associated the New Jerusalem with the hidden spiritual dimensions of London over a hundred a fifty years before the Earthstars discovery came to light. The fact that he lived, for a considerable part of his life, within walking distance of the Earthstars' centre may have some bearing on this matter (See the post-script on Blake's Jerusalem which has been added at the end of the book)

Nevertheless, his references to London as the New Jerusalem are a startling confirmation of the Earthstars discovery and a reminder that it has to be appreciated primarily from a visionary perspective.

Although most of us are oblivious to the Visionary Landscape, it is all around us. As the Irish seer known as AE describes it in his book CANDLE OF VISION, it is; **"The golden age that is about us still for those that have eyes to see."** AE, like Charles Williams, was another member of The Hermetic Order of the Golden dawn and the Golden Age to which he refers, is not just in the past, it is in a different dimension, a perfected state, co-existent with this one. In this sense, the idea of its return is dependant on a change in human perception as much as any change affecting the world around us. New Age sources explain this as a transformation from three dimensional reality to fourth or fifth dimensional existence.

A simple change in how we perceive the world around us might be all that is needed to see our role here on Earth in a new light and precipitate a whole new era of human evolutionary progress in full awareness of our partnership with the formative powers of the living holistic universe.

For those of us who live in densely populated, heavily built-up areas, even rare glimpses of the visionary landscape do not come easy but they may be more likely within the special atmospheres of our sacred sites. After all, these have always been places of power, of vision, of inspiration.

If Sir George Trevelyan's vision and my own dream are to be believed, the process of healing the wasteland has already begun and a transformational energy of unusual intensity has started to activate through the Earth's sacred sites as part of a natural cycle.

As natural spiritual power stations, sacred sites would understandably be the first places to feel its effects. This force, whether you call it the Holy Spirit or an evolutionary impulse, may be barely discernable to the general public, but could rapidly be returning to the higher levels of a former age. If that is correct the reality of these forces may soon be more readily apparent to all.

Many people appreciate that the atmosphere at our sacred sites has the capacity to uplift spirits or simply to induce a sense of peace and tranquility. Whatever the cause, these places possess a natural and potent feel-good factor.

I can personally testify that they are also capable of initiating profound experiences which can help restore our personal connection to the divine and considerably more evidence is available from the records of Dr.

James Swan whose research confirms the effects noticed at sacred sites include dream-like revelations, religious visions, mystical insights and even hearing voices, music or songs.

Those who dismiss such things lightly, ignore overwhelming evidence that this is a very real phenomenon with ability to alter our conscious perception of reality.

That alone can change our world dramatically.

THE EARTHSTARS VISION

There is a tendency to assume that our ancient sacred sites, stone circles, old churches, henge monuments, sacred hills, mounds and groves, are only relevant to the past. Yet our ancestors selected these locations because they were focal points for the Universal Life Force. They link us to the formative forces of creation and evolution, to our origins and our destiny. They mark places of power, places of vision, places of healing, places of inspiration. As such they are far more relevant to our present and future than our past.

The Earthstars sites in particular seem to carry a message very definitely concerned with our future. The network of perfect geometry connecting them was discovered following visions of a white lady figure similar to apparitions of the Virgin Mary. A dream identified the sites as places of spiritual power which may be symbolised as the Waters of Life and very clearly indicated that its purpose was to cleanse, heal and purify the Earth.

Astonishing though this may be, all the subsequent research actually confirmed the message of the dream from a number of other perspectives. The harmony of the Earthstars geometry, for instance, directly reflects the harmony of the forces inherent within it. These patterns are very definitely a harmonious influence and a power for good. The Waters of Life, whether in a dream or anywhere else, symbolise a particularly potent aspect of the Universal Life Force. A regenerative, healing influence. The power to heal the wasteland.

The individual sites, the geometric patterns they form, and the forces inherent within them, seem to be part of a process that will re-attune ourselves, our environment and our planet to the greater harmony of the universe and the cosmic pulse of creation.

This is what Sir George Trevelyan describes as the redemption of the fallen planet in his book OPERATION REDEMPTION. We can ignore this evidence and perpetuate the wasteland mentality or we can use the sites, the energies and the geometry to help transform our world into something considerably better.

Through this geometry, all things are connected and brought into harmony. By recognising the significance of these specific patterns and the laws of harmonic proportion they represent, we can learn to work with them, to acknowledge the divine plan within the material world and begin to build in harmony with nature and the creator, rather than on the whim of this year's designer trends.

In this sense, the Earthstars patterns are a vision to be fulfilled. They are keys to the unity of the holistic universe. They are keys we can use to re-discover our place within that unity, as individuals and collectively, as a planetary populace. They are keys we can use to open the portals to our spiritual dimensions whatever our religion.

Some of the keys are musical as well as geometric. Of particular importance is the significance of the relationship between the 20 and 30 point stars whose harmonious geometry is echoed in musical harmony. 20:30 mirrors the 2:3 ratio of a harmonic fifth and is directly related to the pentagonal geometry.

By asking architects to incorporate these specific geometric harmonies into the proportions of our buildings as some of the classical architects did centuries ago, we can create structures with a spiritual dimension: Structures with the ability to inspire and uplift the spirit of anyone who enters them. If that only results in rooms and buildings designed in golden mean rectangles instead of haphazard proportions, it will at least be a step in the right direction.

By applying similar principles to town planning, to the redevelopment of our cities and to the landscape at large, we can literally build in a community spirit which is both in harmony with the natural order and with its inhabitants. Re-discovering the spiritual energies of the landscape and working with them, instead of ignoring or disrupting them, can create a much better environment, and I would personally be happy to work with any local authority or Diocese on any community projects with these aims.

These ancient spiritual sciences were known and used by our ancestors who built the pyramids, Stonehenge and the great temples and cathedrals of Europe. The knowledge has not been totally lost. It is all around us. It is a living tradition. We need to re-discover how to use it. More particularly, our religions need to re-discover how to use it.

The manipulation of these forces for the benefit of the community was an integral part of ritual and ceremony in past cultures and is as relevant to our spiritual wellbeing today as it was then. The Church of England owns many of churches built upon ancient sites which are natural spiritual power stations. They could very easily re-introduce an awareness of this science of the Holy Spirit into church activities and services. Sound, music, dance, prayer and meditation can all be used to amplify and direct these forces to create a very potent and beneficial effect on the surrounding area and those who live within their sphere of influence. Many of these practices are already part of normal church activities. Building Blake's vision of the New Jerusalem on England's green and pleasant land demands a little more effort, but is not beyond the realms of possibility

The sacred ground plan already exists within nature, for those who care to look for it. It is, in part, evident in the location of some of our churches, sacred hills, ancient monuments and other special places.

Although incomplete, the pattern itself suggests it has the ability to generate a more spiritual atmosphere and a change in the emotional climate over a widespread area. What would happen if the design was completed, the vision fulfilled ?

There are enough vital spots within the Earthstars designs which have not yet been built upon and could be reserved for monuments, public buildings or structures which enhance this greater plan designed to bring the world of humanity into harmony with the structures of the natural world and the universe at large.

Erecting mark stones, fountains, peace gardens, statues of religious figures, stone circles, classical temple follies, sculptures or other structures incorporating the holistic geometry of the Earthstars design can have an enormously beneficial effect on the energies and atmosphere of an area if they are correctly located. The more the plan is established in the real world, the greater its power.

The ground plan has already been laid. Not by me. Not by any human hand. But by the forces of creation, the formative forces of the Universe. It presents a unique opportunity for humanity to accept our role as co-creators of our planet with a higher degree of awareness and to fulfill our true destiny as the children of God.

It is my firm belief that the Earthstars discovery is of global significance and is directly related to the spiritual destiny of Britain.

These have always been very special islands although much of our early history seems to have been deliberately obscured.

As an example, if we are to believe what we are taught in our schools, British history only began with the civilising influence of the Roman invasion of 55 B.C. and prior to that, Britons were all woad-splattered savages. This, of course is the Roman version of events. The British and Welsh records paint an entirely different picture.

What we are not generally told about the Roman invasion of 55 B.C. is that it failed miserably. Caesar and the might of the Roman army, which had subjugated the rest of Europe, were dealt a military defeat of major proportions by the British forces under the command of the British leader, Caswallon. They retreated back to the safety of France 55 days after landing on the Kent coast.

To understand the magnitude of this achievement, we should be reminded that Caswallon's Britons were engaging possibly one of the ablest generals of antiquity, heading an army to which France, Spain, Germany, Africa, Egypt, and parts of Asia all succumbed.

The second attempted invasion of Britain followed less than a year later, in May 54 B.C. when Caesar returned with two divisions of Rome's best fighting troops. This time, the army that had marauded unchecked through Africa and Europe fought for four months and only got as far as St. Albans before being pinned down as they had been the previous year in Kent. A hasty peace was concluded. The Romans returned to France and were wary enough of the British not to mount another attack for almost a hundred years.

Although Caesar depicted his British adversaries as barbarians, other evidence suggests that Britain possessed an ordered civilisation equal to that of Rome. As far back as 400 B.C. in the reign of Molmutius, 16th King of the Britons, the first published British Laws and statements of civil liberty may be found. Amongst other things, they assert that it is the birthright of every Briton to have equality of rights, to be free to come and go as they please and to be protected by their sovereign and country. Molmutius is also said to have built the first roads, for the ancient Britons were skilled charioteers and this played an important part in routing the Roman legions.

By all accounts, Britain was also a centre of great learning, possessing Druid Colleges or Universities at York, Canterbury, Winchester, St Albans, Carlisle, Cambridge, Manchester, Colchester, Worcester, Chester, Cirencester, Dorchester, Carmarthen, Caernarvon, Exeter and Bath. Indeed, many well-known public schools seem to have been founded on or near very ancient sites and may actually mark the locations of Druid colleges.

It is well documented that the Druidic seats of learning were held in high esteem throughout the known world and that Druids and nobles from Europe and elsewhere came here for their education.

R.W. Morgan in ST. PAUL IN BRITAIN writes of The British Druids. **"The ramifications of druidism penetrated indeed into Italy, Greece and Asia Minor; nor did Plato hesitate to affirm that all the streams of Greek philosophy were to be traced, not to Egypt, but to the fountains of the West; The pre-historic poets of Greece, anterior to the mythological creations of Homer and Hesiod were, as their names imply, Druids."**

More important is the fact that Britain seems to have been the first place to accept the Christian religion. Cardinal Pole made this statement at Westminster in 1555.

"The see apostolic from whence I come hath a special respect to this realm above all others, and not without cause, seeing that God Himself, as it were, by providence hath given to this realm prerogative of nobility above all others, which to make plain unto you, it is considered that this island, first of all islands, received the light of Christ's religion"

He is referring to that fact that one of the first Christian churches in the world was founded here at a time when Rome was still throwing Christians to the lions for public entertainment. This was the Church of St. Mary the Virgin at Glastonbury, now the site of the Lady Chapel at the western end of Glastonbury Abbey. Legend tells us it was founded AD 36 - 39 by Joseph of Arimathea, Christ's uncle, who travelled to Britain from The Holy Land via France with twelve companions shortly after the crucifixion.

According to many sources, this is no legend. It is confirmed by no other authority than the church of Rome who might actually have been expected to dispute the fact. St. Augustine himself wrote in a letter to The Pope:

"In the western confines of Britain there is a certain royal island of large extent, surrounded by water, abounding in all the beauties of nature and necessarys of life. In it, the first enophites of the catholic law, God beforehand aquainted them, found a church constructed by no human art, but by the hand of Christ himself, for in the salvation of His people, the Almighty has made it manifest by many miracles and mysterious visitations that He continues to watch over it as sacred to Himself and to Mary the Mother of God."

This confirms that a Christian church already existed in Britain when Augustine arrived, although his understanding is that it was built by Christ himself, rather than by Joseph of Arimathea. When the antiquity of the British church was challenged by the ambassadors of France and Spain before the Roman Catholic Council of Pisa (AD1417), the British delegates

Robert Hallam, Bishop of Salisbury, Henry Chichele (a former Archbishop of Canterbury) and Thomas Chillendon, won the day and the council affirmed that the first Christian church was indeed in Britain. At the Council of Sienna in 1424, the churches of France and Spain had to concede the precedence of the British Church which, it was affirmed, was founded by Joseph of Arimathea immediately after the passion of Christ.

This incredible tale is made more plausible by Cornish and Somerset folklore relating to Joseph's trade as a tin merchant, which had brought him to these islands on many previous occasions. It is widely believed that Jesus and his mother Mary had travelled with Joseph on these visits. Gildas in the sixth century wrote:

"Christ the true sun, afforded his light, the knowledge of his precepts, to this island during the height of the reign of Tiberius Caesar." The height of Tiberius' reign was between AD 20 and 27. If Christ visited Britain with Joseph during those years, it was as a man, not a boy. So perhaps the church at Glastonbury was indeed built by him in those days. William Blake, of course, refers to the legends of Christ's visits to Britain in his patriotic poem Jerusalem.

And did those feet in ancient time walk upon England's mountains green? And was the holy lamb of God on England's pleasant pastures seen ?

And did the countenace divine shine forth upon our clouded hills? And was Jerusalem builded here among these dark satanic mills?

Bring me my bow of burning gold Bring me my arrows of desire. Bring me my spear, O. clouds unfold Bring me my chariot of fire.

I will not cease from mental fight, Nor shall my sword sleep in my hand, Till we have built Jerusalem, in England's green and pleasant land.

As a child, I always found that the words of this hymn created an inexplicably emotional response in me whenever it was included in our school's morning assembly I now know why

The theme of Blake's poem, of course, includes the notion that the New Jerusalem was established here within our hidden spiritual environment and could physically be built in England's green and pleasant land by those with sufficient vision, will and determination.

Not only does the Earthstars discovery confirm Blake's vision, it provides us with the means to make it a reality. Simply by using the Earthstars patterns as a ground-plan or template, we can begin building upon our visionary landscape, rather than building over it.

Since its connections extend to important sacred sites the length and breadth of the land, and beyond, this is not an endeavor relevant just to London. The purpose of the New Jerusalem is **"for the healing of nations"**. The power inherent in the Earthstars discovery can do exactly that. It is a construction of the Universal Life Force; a single unifying web of life and spirit which links all people, all nations, all religions, all life on Earth. It knows no boundaries, or limitations. This is a force with the power to transform ourselves and our world.

Who knows what might happen if the circuitry is completed and connected ?

I, for one, would like to see.

ALWAYS KEEP YOUR FEET

FIRMLY PLANTED IN THE STARS

AND LEAVE THE WORLD

A BETTER PLACE

THAN YOU FOUND IT.

BLAKE'S JERUSALEM
(Notes added to the 2010 edition)

William Blake's visions of Jerusalem within London are extremely relevant to the Earthstars discovery, but to understand why, we must first understand what Blake means by Jerusalem.

Firstly, in his epic poem, he calls Jerusalem **"the emanation of Albion."** Albion is one of the earliest names for the island of Britain. The emanation of Albion means, in essence, the spirit of Britain.

Now, in other traditions, the spirit of Britain is The White Goddess, Bride, or in the Secret Arthurian Tradition, Gwenevere, and Blake is clearly alluding to her. In his illustrations for the poem, he actually depicts Jerusalem as a female figure in white.

To cement the association beyond doubt, he goes on to further describe Jerusalem as both a city and a woman.

This has interesting parallels with the Earthstars discovery and St. John's Vision of the City of Revelation, the New Jerusalem, which he describes as both a celestial city and a woman, in his case, **"prepared as a Bride, adorned for her husband."**

Yet again, there is a clear association with the hidden traditions of the White Goddess, Bride, and where city and woman become one is where the feminine spirit of the land becomes the soul of the city.

Since it relates to our spiritual dimensions and those of the land and planet, it is something only perceived through a visionary experience.

So Blake's visions of Jerusalem, St. John's vision of the City of Revelation and the Earthstars visions may all relate to the same phenomenon, each one seen and interpreted in different ways.

It is a recurring phenomenon of a visionary nature related to the mysteries of the divine feminine, the hidden and occult element of the holy trinity and the elusive, yet ever-present, life force of the planet.

Why Blake chose to present her by the name Jerusalem, is open to debate and I leave that question in the hands of scholars more knowledgeable than myself to answer.

BIBLIOGRAPHY

A.E. **The Candle of Vision.** Prism Press. 1990. First published 1918.
Frederick Bligh Bond; **An Architectural handbook of Glastonbury Abbey.** First Published 1909. Thorsons edition 1981.
Ken Carey; **The Starseed Transmissions.** Starseed Publishing, 1982.
Ken Carey; **Terra Christa.** Uni-Sun. 1985.
Phillip Carr Gomm; **The Druid Way.** Element Books 1993.
Bruce Cathie; **Harmonic 288, The Pulse of the Universe.** Sphere Bks. 1977.
Bruce Cathie; **Harmonic 695; The Energy Grid.** America West Publishers, 1990.
Andrew Collins; **The Seventh Sword.**
Paul Devereux; **Symbolic Landscapes.** Gothic Image, 1992.
Paul Devereux; **Places of Power.** Blandford , 1990.
Paul Devereux; **Earth Memory.** W. Foulsham, 1991.
Paul Devereux; **Earthmind.** Destiny Books, Vermont. 1989.
Paul Devereux; **The New Ley Hunters Guide.** Gothic Image, 1994.
Michael S. Durham; **Miracles of Mary.** Harper Collins.
David Furlong; **The Keys to the Temple.** Piatkus, 1997.
Robert Graves; **The White** Goddess. Faber and Faber 1961.
Joy Hancox; **The Byrom Collection.** Jonathon Cape, 1992.
H.E. Huntley; **The Divine Proportion.** Dover Publications 1970.
K. Jones; The **Ancient British Goddess.** Ariadne Publications
Gareth Knight; **The secret Tradition in Arthurian Legend.** Aquarian Press, 1983.
Gareth Knight; **The Rose Cross and The Goddess.** Aquarian Press, 1985.
Robert Knight and Christopher Lomas; **The Second Messiah.** Century London.
T.C. Lethbridge; **Ghost and Divining Rod.** Routledge and Kegan Paul,
Henry Lincoln; **The Holy Place.** Jonothan Cape, 1991.
Henry Lincoln; **Key to the Sacred Pattern.** The Windrush Press, 1997.
John Matthews and Chesca Potter (editors); **The Aquarian Guide to Legendary London.** Aquarian Press, 1990.
Eagle Man Ed McGaa; **Mother Earth Spirituality.** Harpers, San Francisco. 1990.
Eagle Man Ed McGaa; **Rainbow Tribe.** Harpers, San Francisco. 1992.
John Michell; **The New View Over Atlantis.** First published 1969. Thames and Hudson edition 1973.
John Michell; **Twelve Tribe Nations and the Science of Enchanting the Landscape** (Co-authored with Christine Rhone). Thames and Hudson 1994.
John Michell; **The Dimensions of Paradise.** Thames and Hudson, 1988.
John Michell; **New Light on The Ancient Mysteries of Glastonbury.** Gothic Image. 1990.

John Michell; **City of Revelation.** Garnstone Press, 1972.
John Michell; **The Temple at Jerusalem: a Revelation.** Gothic Image, 2000.
John Michell; **At The Centre of The World.** Thames and Hudson. 1994.
John Michell; **Megalithomania.**
John Michell; **The Earth Spirit; Its ways, shrines and mysteries.** Thames and Hudson, 1975.
Ross Nicholls; **The Book of Druidry.** Aquarian Press.1990.
David Pam; **The Story of Enfield Chase.** Enfield Preservation Soc. 1984.
Nigel Pennick; **The Mysteries of Kings College Chapel.** The Aquarian Press, 1974, 78 and 82.
Nigel Pennick; **The Ancient Science of Geomancy.** Thames and Hudson, 1979.
Nigel Pennick; **Sacred Geometry.** Turnstone Press, 1980.
Nigel Pennick; **Earth Harmony.** Century, 1987.
Michael Poynder; **Pi in the Sky.** Rider, 1992
Peter Quiller; **Merlin The Immortal.** Spirit of Celtia, 1984.
Peter Quiller; **Merlin Awakens.** Dragonfly, 1998..
Jean Richer; **Sacred Geography of The Ancient Greeks,** translated by Christine Rhone. State University of New York Press. 1994.
Greg Rigby; **On Earth as it is in Heaven.** Rhaedus Publications, 1996.
Jamie Sams; **Midnight Song.** Bear and Co. 1988.
William Stirling; **The Canon.** Published by The Research Into Lost Knowledge Organisation. Distr; Thorsons.First Published 1897. R.I.L.K.O. edition, 1974.
The Reverend Gordon Strachan; **Christ and The Cosmos.** Labarum Publications. 1985.
James A Swan; **The Power of Place.** Gateway Books. 1993.
James A Swan; **Sacred Places,** Bear and Co. 1990.
James A Swan and Roberta Swan; **Dialogues with the Living Earth.** Quest Books. 1996.
Sir George Trevelyan; **Operation Redemption.** Turnstone Press, 1981.
Sir George Trevelyan; **Summons to a High Crusade.** Findhorn Press, 1986.
Sir George Trevelyan; **A Vision of the Aquarian Age.** Coventure London. 1977/84.
Guy Underwood; **The Pattern of the Past.** Abacus, 1969.
Alfred Watkins; **The Ley Hunter's Manual.** 1983,1985. First published, 1927.
Alfred Watkins; **The Old Straight Track.** First published 1925. Re-printed by Abacus 1970-93.
William Willoya and Vinson Brown; **Warriors of The Rainbow;** Naturegraph Publishers. P.O.Box 1075, Happy Camp, CA. 96039, USA.
David Wood; **Genisis.** The Baton Press, Tunbridge Wells, 1985.
David Wood and Ian Campbell; **Geneset.** Bellevue Books, 1994.
William Wynn Westcott; **The Magical Mason,** edited by R. A. Gilbert. Aquarian Press, 1983.

USEFUL ADDRESSES.

THE COLLEGE OF PSYCHIC STUDIES: 16 Queensbury Place, South Kensington, London SW7 2EB. 020 7 589 3292.

THE RESEARCH INTO LOST KNOWLEDGE ORGANISATION: www.rilko.net

FOUNTAIN INTERNATIONAL: www.fountain-international.org

THE EARTH HEALING DAY NETWORK: www.earthhealing.co.uk

THE ORDER OF BARDS OVATES AND DRUIDS: P.O. Box 1333 Lewes, East Sussex. BN7 3ZG. www.druidry.org

THE ANTIQUARIAN SOCIETY: www.theantiquariansociety.com

EARTHSTARS RESEARCH

For further information on Earthstars, The Visionary Landscape and the author's other work, please visit www.earthstars.co.uk

INDEX

Act of God by Graham Phillips 151
Acton 103
Albert Einstein: 10
Albert Hall 204
Alexandra Palace 88,197
All Saints Church Childs Hill 83
All Saints Church Fulham 102
All Saints Church, Colville Gardens 103
All Saints Isleworth 86
All Saints Kingston on Thames 143
All Souls Langham Place 86,102, 103,190,197
Anima Mundi 170
Ansgar, Staller to the King: 17
Aquarian Guide to Legendary London 77
Arnold Circus 182,184
King Arthur: 17,19
At The Centre of the World by John Michell 223,226
Avebury 103,219

Baal 123
Bardsey Island: 27
Barking , Barking Abbey 130-131,183,185, 186, 240
Barn Hill Wembley 179, 200, 219
Barnet 127,129,137,143,239,242,243
BBC Broadcasting House 190
Beckenham Palace 83
Bellingham Green 123
Bel, Belinus 123
Belmont Stanmore 219
Beryl Bohea Raine: 13
Bethnal Green 103
The Bevan Family: 23
Birmingham 219
Bletchingly Castle Hill Mound: 58
The Book of Druidry by Ross Nicholls 123, 124,150,250
Humphrey De Bohun 280
Henry Bollingbrook 280
Bornholm 219
Black Elk 260
William Blake 292
Colin Bloy 177, 271, 272,273,274
Brean Down 103,105
Brent Cross 83
Bride, Brigid: 26, 27,29,62,96,245,247-251,278,288
The Brill 102
British Museum 103,197
Brixton 247
Brock House 190-194
Brockwell Park 197
Bromley 83
Brutus: 77
Bryon Court School, Kenton 148
Brynn gwynn: 78,186
Bulstrode Camp: 75
The Byrom Collection by Joy Hancox 168

Cadbury Castle: 14
Caerleon: 14
Caesar's Camp Wimbledon: 68,69,70,92,102,123
Camelot: 14,16, 17, 18,32,245,279

Camlet Moat: 15,19,20,21, 22,27,28,31,33,42,74,97,105,106,109,110,241,277, 278,280,283

Camlet Way: 15,17,20,28,42,143
Camulodunum: 14
Camulus: 17
The Cannon by William Stirling 117,238,245,286
Cathedral Church of the Good Shepherd N16: 102
Chakras 171-173
Chessington 84
Chingford 99,110,143
Circle of the Earth Spirit 124,176,248,250
Circle of Perpetual Choirs 207-210,214
City of Revelation by John Michell 119,126,151,207,208,230,231
The City of Revelation 229-243,257,269
Clay Hill 109
Alan Cleaver 207
Clive Beadon 215-217
Cockfosters: 20
Colchester: 14
The College of Psychic Studies: 12
Andy Collins 210
Keith Critchlow 151,170,285
Crop Circles, 221,222
Croydon: 64,109,112,127,130,137,239
Crystal Palace 197

Dabbs Hill: 75,219
Diana: 2,26
Delphi: 2
The Dimensions of Paradise by John Michell 225
Michael S, Durham: 3
Paul Devereux: 5
Earth Harmony by Nigel Pennick: 27,175
East Ham from Village to County Borough 98
Earthquest News by Andy Collins 210
East Ham Triangle 130,140.144,240
Eltham 110
Enfield 110,143,280
Enfield Weekly Herald: 23
Epping Forest 123,148
Ethelburga 131

Fatima: 3
Feng Shui 175,176
Fermat's Last Theorem by Simon Singh 116
Fibonnacci 114
Fingerprints of the Gods by Graham Hancock 259
Finsbury Park 100,101,197
Robert Fludd 171
Frognal 83,87
Fred Fiske 84
Fountain International 272,,274
Fosse Way: 17
David Furlong 272

Gawina 277
Genisis by David Woods 211,250
Glastonbury: 27,161,163,191-195,207, 297
Glastonbury A Study in Patterns by RILKO 207
Golden Mean 114
Golden Gate 225

Goring on Thames 209-210
Graham Hancock 259
Great Portland street: 47
Great Pyramid 151, 152,189
Green Park: 52,**53,199,201**
Green Knight 277
The Green Stone Martin Keatman and Graham Phillips 283
Grim's Dyke Harrow Weald Common 123
Grimstone's Oak Epping Forest 123
Guilford 162
Guild Hall 199
Gwenevere: 25,30,171,245,277,278
Hamish Miller 274
The Hanwell Triangle 132, 133, 134, 135, 140,144,241
Harrow on the Hill 219
Henry VIII 131

Hermes: 30
The Hermetic Order of the Golden Dawn: 30,281
Hertford Castle: 16
Highbury Hill 197,199
Highgate Hill, 100
The History of Enfield Chase: 15, 18
The History of The Kings of Britain: 16
The History of Tottenham 84
The History of Trent Park: 31
Holland Park 197,199
The Holy Place by Henry Lincoln 214,239
The Holy Blood and Holy Grail 214,247
Holy Spirit/Ghost/Trinity 243,244,245,287
Hopi Prophesy 266 , 267 ,290
Horn Hill 254Horsa: 81
Horsendon Hill Greenford: 68,69,70,72,74,79-81,92,123,179,197,219
Horus: 81
Mr. C. Houston: 17

David Icke: 5
Iona: 27

Jerusalem 223

Kensington High Street 197
The Key to the Ancient Mystery by David Furlong 272
Kings College Chapel 160
The Kingstone Kingston-on-Thames 143,146,182,184
Gareth Knight: 30,32

Lambeth Palace 204
S. Leadbetter Architect 191
The Ley Hunter: 5
Leys: 10
Henry Lincoln 214, 219,239,250
Lindos: 2
LLantwit Major 207
London Stone: 75,77,197,199
Ludgate Hill: 2, 182
Lundy Island 103

Thomas Mallory: 16
Manasara Shilpa Shastra 166,175
Nicholas Mann 226

Sir Geoffrey de Mandeville: 15,16,29,31, 278,279,280
The Magical Mason by William Wynn Westcott 249
Marylebone Basin 191
Mayan Prophesies 259
The Messianic Ley 223
Geoffrey of Monmouth: 16
Shirley McClaine: 5
Ed McGaa: 4,261,263
Medjugorje: 3
Megalithomania: 10
Merlin: 9
Merlin The Immortal by Peter Quiller and Courtney Davis 265,266,284
John Merron 207
The Messianic Legacy 214
John Michell: 3, 4,10,63, 150,175,191,223,224,225,229, 233,238,290
Miller's Well East Ham 98,99,103
Mincheden Oak: 35,37,89
Minchenden Chapel: 35
Movement and Rhythms of the Stars by Joachim Schultz 178
The Mysteries of Britain by Lewis Spence 277
New Jerusalem 229,243,246,250,251,253,257,283,291

Oak Hill Park East Barnet 179,181
Odin 81
The Old Straight Track: 3
On Earth as it is in Heaven by Greg Rigby 218
One Tree Hill Alperton 81,82
One Tree Hill Honour Oak 83
Our Lady of Hal, Arlington Rd: 60
Our Lady of the Annunciation Addiscombe 110

David Pam: 15, 18
Pantheon in Rome 119
Parliament Hill: 35,37.90,146,197
Nigel Pennick: 27,28,41,151,160
Pi in the Sky by Michael Poynder 215-217
Pollard's Hill Norbury: 54-58,64,66,68,69,92,99,112,127,144,148,190,241
Graham Phillips: 16,152
Primrose Hill: 35,37,90,173,181,197
Prittlewell Priory 103
Pythagorus 115,116,151,286

Rainbow Warriors 260,263
Rennes Le Chateau 211-213, 219,239,250
Christine Rhone: 4,223
Richmond Hill 148
Greg Rigby 218
Roman Roads: 72,73, 137
Rottingdean: 59, 60
Russell Hill 58

Sacred Geometry of Washington DC by Nicholas Mann 226
Sacred Geometry Philosophy and Practice by Robert Lawler 120
Sacred Geometry, symbolism and Purpose in Religious Structures by Nigel Pennick 15
St Andrew's Frognal 87,197,199
St. Anne's Dean Street 83
St. Bartholomew The Great 199

St. Benet's Church 204
St. Bride's Fleet Street: 27,75,78,90,197,199,248
St. Dunstan's Bellingham Green: 68,69,70,92
St. Dunstans-in-the-West: 75,197,199
St. Eloy's Well Tottenham 85,90
St. Etheldreda's 204
St. Gabriel's Wanstead: 68-70, 86,89,92,123
St. George's Nine Elms 87,197,199
St. James' Friern Barnet: 45
St. James' Piccadilly 204, 206
St. John's Aldenham 83
St. John's Clay Hill 45,
St. John's Croydon: 64,66
St. John's High Barnet: 35,37
St. John's Old Coulsden 58
St, John Fisher Church Kidbrooke 148
St, Joseph's Highgate Hill: 47, 49
St. Leonard's Turner's Hill: 58
St. Leonard's Shoreditch 137
St. Leonard's Streatham: 49, 61
St Margaret's Barking 131
St. Martin-in-the-Fields 197
St. Mary's Addiscombe: 65,66,109
St. Mary's Beddington: 65,66,105,106,109,110,112,127
St, Mary's Chessington 84
St. Mary's Denham 179
St. Mary's East Barnet: 31,33,34,37,39,42,45,47,68,69,90,91,96,99, 144,148,179,181,190,219
St. Mary's East Ham 132
St. Mary's Hanwell 132
St. Mary's Harrow-on-the-Hill 81, 100,180,219,220
St. Mary le-Boltons 102
St. Mary's Monken Hadley:15,27,42,45,109
St. Mary's Norwood Green 132
St. Mary's Stoke Newington 102
St. Michael's Mount: 27
St, Nicholas, The Bury Stevenage: 58
St. Pancras Church 102
St. Paul's Cathedral: 2,182,201,204
St. Paul's Portland Chapel 191-194
St. Peter's De Vere Street 102
St. Vedast 182
Sacred Places by Dr J. Swan: 4
Jamie Sams: 4
Sir Phillip Sassoon: 23,27,31
The Secret Tradition in Arthurian Legend by Gareth Knight: 30,32
Seething Wells Esher 123
Joachim Shultz 178
Silbury Hill 58,103,104
Sion Park 86,92,96
Sophia: 31,288
Bernadette Soubirou: 3
Southgate 179
Southwark Cathedral 131,186,197
Lewis Spence 278

The Spiritualist Association: 11, 12,13
The Squared Circle 119-129
Star Wars: 13
Stations of the Sun by Prof.Ronald Hutton: 62
Stonehenge: 9,10,153-158,181,186,189,207,221,222
Gordon Stracham 269
Strawberry Hill Twickenham 143
Summons to a High Crusade by Sir George Trevelyan 264
The Sun and The Serpent by Hamish Miller and Paul Broadhurst 274
Dr. James Swan: 4,6

Taliesin in The School of the Poets: 3-
Tarot: 11,26,39,45,80,269,280
Templars: 31,278,280
The Templar Revelations by Clive Prince and Lynn Licknett 247
Temple Church 204,278
Temple Mills 86
Professor Alexander Thom 151
Time Stands Still by Prof. Keith Critchlow 166
Tintagel: 14
Tottenham 84,85
Tower Hill: 78,87,88,90,131,186,197,199
Trent Park: 20,23
Sir George Trevelyan 262,263,268,292,293
Trafalgar Square 83
Twelve Tribe Nations by John Michell: 4,223

Valentine Park 87
Vesica Psics 117
Victoria and Albert Museum 102,204
Vision of the Aquarian Age, by Sir George Trevelyan 262

Waldron Abbey: 29
Walsingham: 3
Washington 226
Wanstead Park 85 Alfred Watkins: 3,43
Watling Park 83,92,96
West Ealing 103
Westminster Cathedral: 55,56,61,198
Westminster Abbey: 55,56,146,163,182 185,186,197,199,201,203
Whitby Abbey: 58
The White Goddess: 13,25, 27,30,32,33,245, 288,247,248,278,279,283,284
Willesden Green: 3
Charles Williams: 30,31,291
Wimbledon Common 123
Robin Winbow: 12
Woodford Green 109
David Woods 211,239,250
The World Soul 170

W.B. Yeats 30

www.ingramcontent.com/pod-product-compliance
Lightning Source LLC
Chambersburg PA
CBHW080845230426
43662CB00013B/2028